Nuclear Engineering Division 2020

Held at the 2020 AIChE Annual Meeting

Online
16 – 20 November 2020

ISBN: 978-1-7138-2308-7

Printed from e-media with permission by:

Curran Associates, Inc.
57 Morehouse Lane
Red Hook, NY 12571

Some format issues inherent in the e-media version may also appear in this print version.

Printed with permission by Curran Associates, Inc. (2021)

For permission requests, please contact AIChE
at the address below.

AIChE
120 Wall Street, FL 23
New York, NY 10005-4020

Phone: (800) 242-4363
Fax: (203) 775-5177

www.aiche.org

Additional copies of this publication are available from:

Curran Associates, Inc.
57 Morehouse Lane
Red Hook, NY 12571 USA
Phone: 845-758-0400
Fax: 845-758-2633
Email: curran@proceedings.com
Web: www.proceedings.com

TABLE OF CONTENTS

604645 Small to Full Height Scale Comparisons of Cesium Ion Exchange Performance with Crystalline Silicotitanate

Abstract:

Crystalline silicotitanate (CST) ion exchanger is planned to be used to remove cesium (^{137}Cs) from the aqueous phase of Hanford tank wastes in preparation for vitrification at the Waste Treatment and Immobilization Plant (WTP). Batch contact testing was conducted to assess cesium exchange kinetics at four different CST particle size distributions. Column testing was conducted at three column sizes, small (2.5% full height), medium (12% full height), and full height, to assess Cs load performance behavior. Testing at the small scale was compared to actual Hanford tank waste testing to verify the validity of simulant tests to reflect full height column performance. CST particle size adjustment was essential in scaling the column dynamics down from full scale to medium and small scales. The mean (d_{50}) CST particle size needed to be reduced from 633 microns to 567 and 542 microns, respectively, to achieve appropriately scaled results. Based on these results, a determination of intraparticle and film diffusion impacts on overall mass transfer coefficients will allow future modeling of the breakthrough performance at a range of process conditions.

606101 Low LET Gamma and High LET Alpha Degradation Studies of the Actinide Lanthanide Separation Process

Abstract:

An organic solution comprising the Actinide Lanthanide Separation Process (ALSEP) extracting solvent consisting of 0.5 M 2-ethylhexyl phosphonic acid mono-2-ethylhexyl ester (HEH[EHP]) and 0.05 M N,N,N',N'-tetra(2-ethylhexyl)diglycolamide (T2EHDGA) in *n*-dodecane were subjected to low LET gamma and high LET alpha irradiation before and after equilibration with 3 M HNO_3. Degradation dose constants revealed greater ligand degradation due to gamma irradiation than alpha irradiation for both ligands and equilibration with nitric acid did not have a significant impact on degradation kinetics for either irradiation source. Identified degradation products were similar for both gamma and alpha irradiation and occurred mostly through the rupture of the $N\text{-}C_{carbonyl}$ and $C\text{-}O_{ether}$ bonds for T2EHDGA and the $C\text{-}O_{ether}$ bond in HEH[EHP], and acid contact appears to alter the degradation pathway by favoring the formation of higher molecular weight recombination products. Mixed T2EHDGA-HEH[EHP]-NO_3 complexes were formed with Nd(III) after extraction from 3 M HNO_3, and low LET gamma irradiation of the Nd(III) loaded organic solution produced similar degradation products as the organic solution absent of Nd(III). Additionally, likely due to the greater radiolytic susceptibility of T2EHDGA than HEH[EHP], a HEH[EHP]-Nd(III) complex appears to form with increasing absorbed dose.

611488 Radiation-Resistant Porous Materials for Capture and Separation of Volatile Radionuclides

Abstract:

The release of volatile radionuclides, which must be captured and subsequently stored, is a major problem for the recycling of used nuclear fuel. Solid adsorbents, in particular ultra-microporous metal-organic frameworks (MOFs), could be efficient in capturing these volatile radionuclides, like ^{85}Kr. However, MOFs are found to be more Xe-philic than Kr and to have a similar affinity between Kr and N_2. In addition, the adsorbent needs to have good radiation stability. In order to overcome these challenges, we test a series of ultra-microporous MOFs, SIFSIX-3-M (M= Zn, Cu, Ni, Co or Fe) for their potential in ^{85}Kr separation and storage using a two-bed breakthrough method. These materials were found to have higher Kr/N_2 selectivity than the current benchmark materials, which lead to a notable decrease in the nuclear waste volume. The materials have been systematically studied for gamma and beta irradiation stability, which demonstrates that the metal centre in these isostructural frameworks plays a crucial role in their radiation resistance.

599497 Aging Study of Ag-Exchanged Mordenite and Ag-Functionalized Silica Aerogel in Gas Mixture of Water Vapor and NO_X Streams

Abstract:

Spent nuclear fuel reprocessing processes generates various volatile radioactive species including ^{3}H, ^{14}C, ^{85}Kr and ^{129}I in company with significant amount of water and NO_x vapors into the off-gas streams. ^{129}I is one of those radioactive species which has a long half-life (1.6 x 10^7 years) and negative impacts on the human body and the environment if released to the environment. Many studies have been carried out to capture radioactive iodine from spent nuclear fuel reprocessing off-gas streams. Hydrogen-reduced silver mordenite (Ag^0Z) and silver functionalized silica aerogel (Ag^0-aerogel) have been introduced as solid adsorbents which can effectively capture gaseous iodine. However, the impact of the off-gas streams should be considered as it can result in the decrease of the performance of adsorbents due to the chemical and physical changes to the adsorbents when exposed to the off-gas streams. In previous studies, the aging impacts of off-gas streams on the kinetics of iodine capture on both Ag^0Z and Ag^0-aerogel were investigated. The adsorbents were aged in dry air, humid air, 1% NO and 2% NO_2 respectively for up to 6 months at 100°C ~ 200°C, and iodine adsorption experiments on the aged adsorbents were conducted. Results showed that humid air, 1% NO and 2% NO_2 had negative impacts on the performance of Ag^0Z and Ag^0-aerogel; especially, 2% NO_2 has the most adverse impact on the iodine loading capacity of the both adsorbents. Iodine loading capacities of the aged Ag^0Z and Ag^0-aerogel were decreased with increasing aging time and temperature.

In this study, aging impact of a gas mixture including water, NO and NO_2 on both adsorbents, Ag^0Z and Ag^0-aerogel, was investigated to validate the impacts of each gas stream on the adsorbents. At this writing, aging experiments of Ag^0Z and Ag^0-aerogel in the gas mixture and adsorption experiments on the aged adsorbents are being conducted. In addition to these experiments, chemical and physical changes on the aged adsorbents through aging process will be observed by conducting characterization study with scanning electron microscopy (SEM), X-ray photoelectron spectroscopy (XPS), X-ray powder diffraction (XRD). Also, the kinetic studies of both aged Ag^0Z and Ag^0-aerogel in gas mixture will be discussed.

602684 Adsorption of Organic Iodides from Vessel Off-Gas (VOG) Streams on Silver-Containing Adsorbents

Abstract:

^{129}I is generated in the Vessel Off-Gas (VOG) streams in the organic form such as CH_3I, C_2H_5I, ..., and $C_{12}H_{25}I$. Due to the radioactive nature, the organic iodides must be removed before the streams can be released to the atmosphere. However, comparing with other Off-Gas streams, organic iodides concentration in VOG is extremely low (less than 100 ppbv). Therefore, more specific studies on the adsorption process at low concentrations are required. Multiple materials have been studied in this area including reduced silver exchanged mordenite (Ag^0Z), silver functionalized silica aerogel (Ag^0-Aerogel) and silver nitrate impregnated alumina (AgA). The most efficient adsorbent was preliminarily determined by conducting single-layer low concentration adsorption experiments. Because of the relatively high adsorption rate and resistance to the aging effect, Ag^0-Aerogel was considered as the superior material among Ag^0Z, Ag^0-Aerogel, and AgA.

Furthermore, to understand the adsorption kinetics, 113, 266, 1130 and 10400 ppbv CH_3I adsorption experiments on Ag^0-Aerogel at 150 °C were performed. A core-shrinking process was observed, and the corresponding shrinking core model was applied to determine the reaction rate constant and CH_3I pore diffusivity in Ag^0-Aerogel. A constant pore diffusivity was calculated to be approximately 4.59 $\pm$ 0.10 $\times 10^{-4}$ cm^2/s. The determined reaction rate constants were changing with CH_3I concentration, indicating a possible non-1st order reaction. Therefore, the shrinking core model was modified and we found the reaction order is approximately 1.37. To understand the rate-determine step, the contributions of multiple physical and chemical processes in the adsorption were calculated. The results show that the adsorption process is overall controlled by pore diffusion, but at low concentration, the reaction between CH_3I and silver particles is more significant and the adsorption process can be reduced to a surface reaction at low concentrations in a relatively short period (in at least one year for 113 ppbv compared with equilibrium time of more than 50 years).

596628 Modeling and Simulation of 'fuzz' Formation Onset in Plasma-Facing Materials

Abstract:

Fusion energy is the most promising carbon-free and sustainable energy source for the 21st century and beyond. However, practical realization of economic fusion electricity on the grid is extremely challenging. One particularly challenging problem is the severe surface degradation in the form of nanometer-sized tendrils, or porous “fuzz”, observed following low-energy helium exposure of plasma-facing-component (PFC) tungsten in linear plasma devices and fusion tokamaks. Tungsten is the chosen PFC material for ITER (International Thermonuclear Experimental Reactor) due to its superior thermomechanical properties and relatively low sputtering yield. Helium produced in the deuterium-tritium fusion reaction is a natural impurity in the plasma and it is extracted from the plasma through the divertor in ITER. Experimental studies have revealed that high particle flux of this low-energy helium produces fuzz-like nanofibers at the tungsten divertor within the temperature range from 900 K to 2000 K (consistent with anticipated heat loads and temperatures expected at the divertor wall), which has adverse effects on the reactor performance.

Developing predictive models capable of accessing the spatiotemporal scales relevant to the fuzz formation process is essential for understanding the growth of such complex surface features. Here, we focus on the development of an atomistically-informed continuous-domain model, the outcome of a hierarchical multiscale modeling strategy over disparate spatiotemporal scales, for simulating the onset of fuzz formation in He plasma-irradiated tungsten and validating the model by comparing the simulation predictions with measurements from carefully designed experiments. The model accounts for the development of stress in the He-implanted tungsten layer due to formation of over-pressurized He bubbles in the layer, with large-scale MD simulation results used to parameterize the corresponding constitutive equations. Based on this model, we have conducted self-consistent numerical simulations of the He-implanted tungsten surface morphological evolution and compared the simulation results with experimental measurements. We have also carried out a parametric sensitivity analysis for the effects of the He concentration in the He-implanted tungsten and the He nanobubble size on our model predictions.

Another critical aspect of PFCs is their response to fluctuating heat loads, which can have a significant effect on the fuzz formation. We report a systematic study of the PFC tungsten surface morphological response over a temperature range between 753 K and 933 K to understand the temperature dependence of fuzz formation, under conditions similar to those at the divertor of fusion reactors. Temperature influences the onset of fuzz formation through the strongly temperature-dependent pressure of subsurface helium bubbles formed by implantation and surface diffusivity of tungsten, in addition to the thermal softening of the elastic moduli of tungsten. We explore the effects of helium concentration in the implanted tungsten and helium bubble size, report surface response diagrams predicted over the above temperature range, and interpret the trends exhibited in such diagrams. Our findings elucidate the role of plasma-irradiation duration in fuzz growth and demonstrate the possibility of fuzz formation at temperatures well below the range explored in experimental studies.

Furthermore, we present a cluster-dynamics model to simulate the sub-surface helium bubble evolution, which plays an important role in the surface morphological dynamics of He-implanted tungsten. Our cluster-dynamics model, implemented numerically in our code Xolotl, has been benchmarked based on large-scale MD simulation results and experimental measurements. For realistic predictions of the sub-surface He concentration, a simplified helium bubble bursting model is used in Xolotl to take into account the gas release occurring when a bubble is near the tungsten surface. Incorporating additional relevant physical processes into future versions of the model of PFC surface morphological response also is discussed.

597414 Elastic Properties of Plasma-Exposed Tungsten Predicted By Molecular-Dynamics Simulations

Abstract:

Studying the impact of helium (He) ion implantation on the mechanical properties of tungsten is of utmost importance for evaluating tungsten as a plasma-facing component (PFC) in nuclear fusion devices. Toward this goal, we have performed systematic molecular-dynamics computations of the elastic properties of single-crystalline tungsten containing regular periodic arrangements of structural defects, voids and over-pressurized He nanobubbles, related to plasma exposure of PFC tungsten in nuclear fusion devices. Our computations reveal that the empty voids are centers of dilatation resulting in development of tensile stress in the tungsten matrix, whereas He-filled voids (nanobubbles) introduce compressive stress in the plasma-exposed tungsten. We find that the dependence of the elastic moduli of plasma-exposed tungsten, namely, the bulk, Young, and shear modulus, on its void fraction follows a universal exponential scaling relation. We also find that the elastic moduli of plasma-exposed tungsten soften substantially as a function of He content in the tungsten matrix, following an exponential scaling relation; this He-induced exponential softening is in addition to the softening caused in the matrix with increasing temperature. A systematic characterization of the dependence of the elastic moduli on the He bubble size reveals that He bubble growth affects significantly both the bulk modulus and the Poisson ratio of plasma-exposed tungsten, while its effect on the Young and shear moduli of the plasma-exposed material is weak.

Having established that the elastic properties of PFC tungsten depend on porosity, corresponding to the volume fraction of He-filled bubbles in the tungsten matrix, and the He bubble size, we expect that the scaling relations derived for the elastic moduli as a function of porosity and bubble size will be valid for the actual near-surface region of plasma-exposed tungsten, which contains spatially random distributions of irregularly shaped bubbles with a narrow size distribution. Specifically, we expect our findings based on a regular bubble array in tungsten to be in semi-quantitative agreement with the elastic properties of actual PFC tungsten to within the statistical errors related to the actual bubble size and spatial distribution. Our findings contribute significantly to our fundamental understanding of the mechanical response of PFC materials upon plasma exposure as well as to the development of a structure-properties database in support of predictive modeling of the dynamical response of PFCs in nuclear fusion devices.

601252 Thermal Gradient Effect on Helium and Intrinsic Defect Transport in Tungsten

Abstract:

Plasma-facing materials (PFMs) in nuclear fusion reactors are expected to withstand stringent conditions, with high heat and particle fluxes that modify the materials microstructure. These fluxes can create strong gradients of temperature and concentration of diverse species in the PFM. In addition to the helium (He) ash and hydrogenic species, neutron particles generated in the fusion reaction will be implanted in the material creating intrinsic point defects, such as vacancies and self-interstitials atoms (SIAs), and their clusters. Additional vacancies and SIAs are generated by trap mutation reactions undergone by helium clusters in the PFM bulk and in regions near the plasma-exposed surfaces. These defects as well as He atoms and small mobile helium clusters will then migrate in the presence of the afore-mentioned gradients.

In this work, we use nonequilibrium molecular-dynamics (NEMD) simulations to study the transport of He, vacancies, and SIAs in the presence of a thermal gradient in tungsten used as PFM. We find that, in all cases, the intrinsic point defects and impurity atoms tend to migrate toward the hot regions of the material and calculate their concentration profiles in the direction of the temperature gradient. We also analyze thermal and species transport in tungsten within the framework of irreversible thermodynamics. The resulting concentration profiles from the NEMD simulations are in agreement with the predictions of irreversible thermodynamics. We compute a negative heat of transport for each species analyzed, which indicates that the respective driven species fluxes are directed opposite to the heat flux. These results have important implications for PFMs in fusion environments, mostly when abnormal operation (edge-localized modes or disruptions) occurs in the plasma, which increases the heat flux toward the material and intensifies the thermal gradients. We demonstrate that when thermomigration, i.e., drift species transport driven by the thermal gradient also known as Soret effect, is considered, the resulting steady-state profiles differ significantly from those when species transport is decoupled from heat transport.

597997 Effects of Competing Kinetics and Material Property Anisotropies on Surface Morphological Evolution in Plasma-Facing Tungsten

Abstract:

Tungsten (W) is the chosen plasma-facing component (PFC) material in the International Thermonuclear Experimental Reactor (ITER) due to its superior thermomechanical properties and relatively low sputtering yield. A large body of experimental evidence has established that PFC tungsten suffers severe surface degradation as a result of exposure to high fluxes of helium (He) and extreme heat loads. Specifically, experiments have shown that high particle fluxes of low-energy helium implantation in tungsten within the temperature range from 900 K to 2000 K is responsible for the formation of the so-called "fuzz" nanostructure, which consists of fragile nanometer-sized tendrils. The formation of such surface nanostructure has adverse effects on the mechanical behavior and structural response of PFC tungsten as well as on the reactor performance.

Here, we report results on the surface morphological evolution of PFC tungsten and examine a number of factors that impact such evolution. Our analysis is based on self-consistent dynamical simulations according to an atomistically-informed continuous-domain surface evolution model that has been developed following a hierarchical multiscale modeling strategy and can access the spatiotemporal scales of relevance to fuzz formation. The model accounts for PFC surface diffusion driven by the compressive stress originated from the over-pressurized He bubbles in a thin nanobubble region, which forms in the near-surface region of PFC tungsten as a result of helium implantation, in conjunction with formation of self-interstitial atoms in tungsten that diffuse toward the surface. The model also accounts for the softening of the elastic moduli of PFC tungsten, both thermal softening at high temperature and softening due to He accumulation in tungsten upon implantation, and the elastic moduli are time dependent and evolve until the He content in tungsten reaches its saturation level. We explore the competing kinetics between surface diffusion processes, the rates of which vary with varying surface crystallographic orientation and temperature, and of helium accumulation to a saturation level in the irradiated tungsten, which have significant effects on the onset of fuzz formation. We find that the characteristic time scale for He accumulation may be comparable to the characteristic time scale for stress-driven surface diffusion on surfaces with low activation barriers for diffusion, such as the W(110) surface, at high temperatures that further accelerate the surface diffusion rates; the former time scale is controlled by the He flux in the experiment and the material temperature for low-energy implantation. Such competition is crucial in determining the plasma exposure period required to lead to nanotendril growth.

Furthermore, we explore the effects on the resulting PFC surface morphology of anisotropy in the PFC material properties, including crystal elastic anisotropy, surface free energy anisotropy, and surface diffusional anisotropy, all of which play a role in determining the geometrical aspects of the features appearing in the evolving surface morphology of He-implanted tungsten under realistic irradiation conditions. All of these material property anisotropies are properly incorporated into our PFC surface evolution model and improve its predictive capabilities. The model is further validated by comparisons of its predictions with detailed PFC surface characterization experiments.

604140 Minimizing Surface Turbulence in Casting

Abstract:

Surface turbulence is a leading cause of defects in casting. This turbulence is distinct from traditional turbulence that occurs in bulk fluids and indicates the increased surface area that develops from wavy fluid surfaces and droplets pinching off. In liquid metal at elevated temperature, in contrast to water, these surface react (e.g., forming oxides or carbides) to form thin surface layers that invaginate or pair to bilayers. These bilayers are recognized as a leading source of bulk defects that decrease material properties of metals formed by casting processes. In this talk, we explore staggered posts along the edges of tilt cast molds as a means of reducing surface turbulence. The development of surface turbulence is controlled by the Weber number, the product of fluid density, a length scale (here the width of the mold), and local fluid velocity divided by the surface tension. We find that surface turbulence develops when the Weber number exceeds six. The posts reduce the Weber number by attenuating the velocity more so than the side runner also explored. The presentation highlights development of a 3D printed test platform, computational fluid dynamics modeling of the tilt casting process, and quantitative comparison of their results.

Savannah River
National Laboratory®

The Nuclear Engineering Division, Robert E. Wilson, and the impact of chemical engineers on nuclear energy

2020 Robert E. Wilson Award address

Dr. Maximilian (Max) B. Gorensek, PE
Advanced Modeling, Simulation, and Analytics

2020 Virtual AIChE Annual Meeting
Tuesday, November 17, 2020

The Nuclear Engineering Division

NED
Nuclear Engineering Division
An AIChE Community

- **First and oldest Division organized within AIChE**
  - 67-year history
- **Grew out of Nuclear Engineering Committee of late 1940s**
  - Chaired first by Miles C. Leverett, '76 then John R. Hoffman, '77
- **Established December 1953**
  - 1954 according to NED's "about" page
  - Sponsored First International Conference on Nuclear Engineering, Ann Arbor, MI, June 1954
    - *107 papers, 1200 attendees*
- **First professional organization for then nascent nuclear industry**
  - Initially open to all disciplines, no prerequisite AIChE membership
- **More than 1,000 members by 1955**
- **90% of NED membership chemical engineers by 1964**
  - Other disciplines developed own professional societies
- **Non-members interested in NED encouraged to join AIChE**

Impact of Atoms for Peace

- **By 1953, nuclear "horse was out of the barn"**
  - Both US and USSR had detonated thermonuclear weapons
- **President Eisenhower gave "Atoms for Peace" address to UN General Assembly December 8, 1953 calling for**
  - Creation of international agreements
  - Development of peaceful applications of nuclear energy and technology
  - Democratization of discussion about nuclear power
- **NED development and growth spurred by Manhattan Project legacy**
  - US nuclear weapons program heavily reliant on chemical engineering innovations
    - *Declassification made new technologies available for civilian use*
    - *NED provided forum for discussion and dissemination*
  - "Atoms for Peace" led to big push for peaceful applications of nuclear energy
  - Promise of power "too cheap to meter"
    - *attributed to Lewis Strauss, then AEC chair in speech to National Association of Science Writers September 16, 1953*

Chemical engineering innovation example – gaseous diffusion uranium isotope separation process

- **Photos show original K-25 Manhattan Project gaseous diffusion plant at Oak Ridge, TN (since demolished)**
- **Exploits difference in mean free speed of gas molecules of different isotopic composition – isotopes pass through diffusion barrier at different rates proportional to mean free speed**
- **Uranium can be converted into UF_6, which is a gas at reasonable temperature and pressure**
- **Separation per stage very small – literally need thousands of stages to achieve HEU**

$$\alpha = \frac{{}^{235}UF_6 \text{ in enriched fraction passed thru membrane}}{{}^{235}UF_6 \text{ in depleted fraction left behind}}$$

$$= \sqrt{\frac{m_{^{238}UF_6}}{m_{^{235}UF_6}}} = \sqrt{\frac{352}{349}} = 1.00429$$

Chemical engineering innovation example – Girdler Sulfide heavy water production process

- About 1 out of every 8,700 atoms of hydrogen is deuterium (based on 2010 IUPAC isotopic abundance compilation)
- Therefore, at least 8,700 gal of natural water are needed to extract 1 gal of heavy water (actual amount depends on how much D remains in tails – depleted water)
- Consequently, heavy water production requires large flows of water (only a very small portion of the Savannah River Plant heavy water production facility shown below)

Chemical engineering innovation example – plutonium uranium reduction extraction

- Plutonium and uranium separated and purified from spent fuel or irradiated targets from nuclear reactors
- Fuel rods dissolved in nitric acid which is subsequently contacted with an organic solvent containing tributyl phosphate, selectively extracting plutonium and uranium
- Process carried out in large, long shielded concrete "canyons"
- Hanford canyon interior shown under construction on the left, completed Savannah River Plant canyon interior on the right

Our Friend the Atom

- **1950s-1960s golden years for nuclear energy**
  - 3 wishes from Our Friend the Atom: power, food and health, peace
  - Rapid growth in commercial nuclear power generation
    - *1st US commercial power reactor on-stream at Shippingport, PA December 1957*
    - *Dresden-1, 1st privately-funded US nuclear plant on-stream October 1959 in Illinois*
    - *3rd US nuclear power plant, Yankee Rowe on-stream in Massachusetts August 1960*
    - *1st US nuclear power plant ordered as economic alternative to fossil-fuel plant December 1963 (Oyster Creek, Jersey Central Power & Light)*
  - Advances in nuclear medicine and radiation therapy
- **Cold War also fed growth in production of special nuclear materials**
  - Savannah River Plant construction authorized by President Truman November 1950
  - Plutonium and tritium production a national priority
- **Large number of chemical engineers employed in nuclear industry**
- **NED leadership in nuclear engineering**
  - Co-sponsored subsequent Nuclear Congresses following 1st in 1954
  - Responsible for ANSI Section N46, Nuclear Fuel Cycle Facilities until 1983

Origins of the Robert E. Wilson Award

- **Idea for divisional award raised several times starting with NED inception**
- **NED Honors and Awards Committee established in 1965**
  - W. Kenneth Davis, *69, chair; S.H. Brown and Bernard Manowitz (then NED chair) members
  - Desire to recognize significant chemical engineering contributions to Manhattan Project continuing into defense and peacetime applications of nuclear energy
- **Committee established basic criteria by late 1966, leaving 2 open questions**
  - Limit award to AIChE members?
    - *Minority opinion against, non-chemical engineers might make significant contributions*
    - *Majority in favor, other awards all required membership (except Professional Progress Award)* ✓
  - Name after Robert E. Wilson or keep generic (e.g., Nuclear Chemical Engineering Award)?
    - *Other eminent chemical engineers' names considered, but many still living in 1966*
    - *Majority of chemical engineering pioneers polled by committee favored Wilson because of his association with overall Manhattan Project* ✓
- **NED ready to seek approval from AIChE Council December 1966**
  - Approached F.J. Van Antwerpen about establishing award for first presentation in 1967
    - *Concern about reluctance to expand number of awards and create new category – Division awards*
    - *Concern about resistance to naming awards for individuals*

On the reason why AIChE membership is a prerequisite for the award

"The story has been told that many years ago at Harvard University there was an event which both the President of Harvard and President Coolidge attended, and the event was duly recorded in a Harvard newspaper, which stated "the President and Mr. Coolidge attended". The point is that frequently many organizations recognize their own, and we of the Nuclear Engineering Division of the American Institute of Chemical Engineers thought it was important that we recognize chemical engineers who have made important contributions to the nuclear energy field."

Attributed to Bernard Manowitz

On the reason why the award is named for Robert E. Wilson

"After a lot of thought and discussion between the Executive Committee and many members of the Nuclear Engineering Division, it was thought that the most appropriate thing to do would be to set up an Award in Nuclear Chemical Engineering. We approached the Council for approval of such an award and the Council approved it. It was entirely fitting, we thought, that the award be named after Robert E. Wilson, the distinguished chemical engineer who made outstanding contributions as the Chairman of the Board of Standard of Indiana and as a Commissioner of the Atomic Energy Commission. That is the background of the Robert E. Wilson Award in Nuclear Chemical Engineering."

Attributed to Bernard Manowitz

Creation of the Robert E. Wilson Award

- **Letter for AIChE Council to F.J. Van Antwerpen from John A. McBride, *91, NED Chairman December 23, 1966**
  - Requested approval of Divisional award to recognize chemical engineers who have made important contributions in nuclear arena
  - Requested approval to name award after Robert E. Wilson
- **Request approved at AIChE Council meeting January 19, 1967**
  - Contingent upon approval by Robert E. Wilson's family (in particular, his widow)
  - AIChE president Theordore A. Burtis to send letter to Wilson's widow on behalf of AIChE and NED
- **Burtis' letter to Mrs. Wilson sent January 31, 1967**
  - NED intended to mail out Call for Nominations early in 1967 with nominations due to Honors and Awards Committee by March 31
  - Unbeknownst to AIChE, Mrs. Wilson had left for an extended trip overseas, not returning home until mid-April
    - *With time running out and no word from family, Honors and Awards Committee considered making award generic*
    - *February NED Newsletter mistakenly sent out with copies of brochure announcing award in Wilson's name, precipitating crisis*
- **Second letter sent to Wilson's daughter March 9, 1967**
  - Favorable response received, request forwarded to stopping point on Wilson's widow's itinerary
  - Mrs. Wilson replied in a letter to the AIChE dated March 27 with her approval
- **First Call for Nominations revised to extend deadline for nominations to April 30**

Original award process

- **Call for Nominations sent to NED membership at beginning of each year**
  - Award criteria
  - Nomination requirements
  - March 31 due date
- **Honors and Awards Committee responsible for managing process**
  - Receive and review nominations
  - Perform initial screening and rank ordering
  - Forward best few nominees to NED Executive Committee
- **NED Executive Committee makes final selection of award winner**
  - Honors and Awards Committee forwarded 3 nominations in 1967, recommending Crawford H. Greenewalt, *67 for inaugural award
  - Greenewalt won distinction of being first Robert E. Wilson awardee
  - Other 2 nominees received Robert E. Wilson award in subsequent years
- **Nominally an annual award**
  - Award need not be made if no suitable candidate nominated
  - To date award has been made 54 consecutive years since 1967

Robert E. Wilson awardees to date

No.	Year	Name	No.	Year	Name	No.	Year	Name
1	1967	Crawford H. Greenewalt	19	1985	James R. Merriman	37	2003	Randall R. Robinson
2	1968	Manson Benedict	20	1986	Alfred Schneider	38	2004	Nathan A. Chipman
3	1969	W. Kenneth Davis	21	1987	Raymond G. Wymer	39	2005	Charles W. Forsberg
4	1970	Robert B. Richards	22	1988	Samuel J. Beard	40	2006	Arlin L. Olson
5	1971	Stephen Lawroski	23	1989	Lawrence T. Lakey	41	2007	William D. (Bill) Rhodes
6	1972	F. C. Culler, Jr.	24	1990	Martin J. Steindler	42	2008	Ralph Leonard
7	1973	Lombard Squires	25	1991	John A. McBride	43	2009	Sharon M. Robinson
8	1974	James A. Buckham	26	1992	Finis S. Patton	44	2010	Jack Watson
9	1975	Milton Levenson	27	1993	John W. Bartlett	45	2011	Terry A. Todd
10	1976	Miles C. Leverett	28	1994	S. George Bankoff	46	2012	William A. Summers
11	1977	John R. Hoffman	29	1995	George R. Jasny	47	2013	Robert T. Jubin
12	1978	Edward A. Mason	30	1996	Hans K. Fauske	48	2014	T. Bond Calloway, Jr.
13	1979	William P. Bebbington	31	1997	Theo G. Theofanous	49	2015	David W. DePaoli
14	1980	Thomas H. Pigford	32	1998	Michael J. Bell	50	2016	Jack D. Law
15	1981	Walton A. Rodger	33	1999	William F. Holcomb	51	2017	Dan Lambert
16	1982	Richard C. Vogel	34	2000	Herbert S. Isbin	52	2018	Robert S. Eby
17	1983	Charles F. Bonilla	35	2001	Glen E. Benedict	53	2019	Paul Taylor
18	1984	Leslie Burris	36	2002	Rowland E. Felt	54	2020	Maximilian B. Gorensek

Who was Robert E. Wilson?

- **Robert E. Wilson made a significant impression on the NED**
  - History suggests overwhelming favorite of Manhattan Project pioneers for award namesake
    - *1983 presentation by F.W. Woodfield states, "because of his strong identification with the overall Manhattan Project program, and not just one of the three major teams on the Project"*
  - Yet no mention in Manhattan Project history
    - *He served as consultant to Petroleum Unit of Office of Production Management from 1940-1941*
    - *Fostered close relationships between Army and Navy and petroleum industry, helped establish petroleum product specifications*
    - *Beginning in 1942 served on four committees of Petroleum Industry War Council (composed of 78 industry executives)*
  - Not clear whether he had any direct role in Manhattan Project
- **Robert E. Wilson had a brief but impactful "second career" in the nuclear industry**
  - Appointed to serve in advisory capacity to Atomic Energy Commission 2 years before retirement from oil industry and 8 years before his death
  - No apparent connection to nuclear energy before that

Robert Erastus Wilson, 1893-1964

- "I have made few outstanding scientific discoveries. My principal contributions to science were probably in the field of generalizing scattered facts, theories, and observations and in applying scientific principles to the solution of practical problems."

- "Four significant facts about my life:

1. I married a secretary in 1916.
2. I hired my first secretary in 1919 (Catherine V. Ogilvie).
3. Both of them are still with me.
4. They are good friends!"

Autobiographical statement to National Academy of Sciences

Early life

- **Born March 19,1893 in Beaver Falls, PA**
  - Eldest of 4 children
  - Mathematics professor father (Geneva College)
- **Moved to Wooster, OH in 1900 when father took position at College of Wooster**
- **Father died in 1907 when Wilson was 14**
  - Mother set up boarding house for college students
  - All 4 children pitched in to help
  - All 4 children graduated from college, 2 went on to graduate school by dint of work and scholarships
- **Graduated magna cum laude from College of Wooster in 1914 with a Ph.B. degree**
  - As had his father in 1889 and later a daughter in 1943
- **Married Pearl M. Rockfellow in 1916**
  - Parents of 3 daughters
- **Enrolled at MIT, earning his BS in chemical engineering in 1916**
  - Stayed on as research associate in the Research Laboratory of Applied Chemistry under William H. Walker

Early career

- **Commissioned as a captain in the Army with outbreak of WWI**
  - Directed Chemical Warfare Service R&D
  - Promoted together with James B. Conant (later president of Harvard) to major, both at age 25 (youngest then in service)
  - Helped create more efficient gas absorbents including soda limes and impregnated charcoals for gas masks
- **Returned to MIT in 1919**
  - Director, Research Laboratory of Applied Chemistry and associate professor of chemical engineering
  - Consultant to Arthur D. Little, Inc.
  - Published research on mechanism of corrosion of iron, mechanism of lubrication, and flow of fluids through pipelines
  - Developed accurate methods of measuring effective volatility of motor fuels
- **Left academia in 1922**
  - Recruited by Standard Oil Indiana to serve as assistant director of research
    - *Asked for 40% increase in salary (to $14,000/year) expecting to be turned down*
    - *Standard Oil Indiana's chairman initially balked at such a high salary for a 29-year old*
    - *Talking with Wilson changed his mind and Wilson was hired*
  - Robert E. Wilson remained with Standard Oil Indiana for 36 years

Standard Oil Indiana (1922-1958)

- **Areas of expertise in petroleum refining included**
  - Methods of reducing evaporation losses in storage
  - Improvements in cracking
  - Coking of residual fuels by delayed coking process
  - Fundamental data concerning
    - *properties of petroleum hydrocarbons*
    - *solvent extraction of lubricating oils*
    - *use of propane as refining agent for separation of wax*
    - *separation of asphalt from heavier fractions of petroleum*
- **Inventor or co-inventor of 89 US and 14 foreign patents**
- **Author or co-author of over 100 technical publications**
- **Steady progression and growth in technical leadership**
  - Head of the Development and Patent Department
  - Membership on Board of Directors
  - Director of Research
- **Named vice chairman of principal subsidiary, Pan American Petroleum and Transport Co. in 1934**
  - Later became president
- **Promoted to Chairman of the Board and Chief Executive Officer of Standard Oil Indiana in 1945, serving for 13 years**
  - Company doubled net worth under his tenure
  - Standard Oil Indiana became one of 10 largest US firms

Recognized as an eminent US chemical engineer

- Chemical Industry Medal 1939
- Perkin Medal, Society of Chemical Industry 1943
- Lord Cadman Memorial Medal, British Institute of Petroleum 1951
- Northwestern University Centennial Award 1951
- Elected to National Academy of Sciences 1954
- Washington Award, Western Society of Engineers 1956

- 18 honorary degrees

- NAS biography does not show an ScD from MIT, but known as "Dr. Wilson"

Active in technical societies

- Alpha Chi Sigma
- American Chemical Society
  - Chairman of Division of Physical Chemistry
  - Chairman of Division of Industrial and Engineering Chemistry
  - Director
- American Institute of Chemical Engineers
- American Nuclear Society
- American Philosophical Society
- American Society for Testing and Materials
- National Academy of Sciences
- Society of Automotive Engineers
  - Director

Active in public service

- Life member of Corporation of Massachusetts Institute of Technology
- Trustee of University of Chicago
- Chairman of the board, College of Wooster
- Established Amoco Foundation
  - Dedicated to the aid of educational and other public institutions
- Author of scores of articles for a wide range of publications, including Saturday Evening Post
- Numerous appearances on radio and television
- Delivered more than 500 public addresses on topics ranging from atomic energy to religion
  - Frequent quip: "Among businessmen I pose as a scientist; among scientists, as a businessman."
  - Another quote: "Most scientists, as they learn more about the wonders of nature, grow in respect for the Creator, many of whose wonders they are barely beginning to understand, let alone duplicate."
  - Would often compress man's 500,000 years of development into 50 years to illustrate recent progress
    - *First printing press only 2 weeks ago*
    - *Only within last day did he have radio, television, rayon, nylon, sulfa drugs, and 100-octane gasoline*

Brief but impactful involvement with nuclear energy

- In 1956, 2 years before his retirement from Standard Oil Indiana, President Eisenhower appointed him to the 9-man General Advisory Committee of the US Atomic Energy Commission (AEC)
- 4 years later, in1960, President Eisenhower named him one of 5 US AEC commissioners
  - Led successful effort to amend Atomic Energy Act to permit private ownership of fissile material
  - Stimulated expansion of US nuclear power generating capacity
  - Interested in use of nuclear power as instrument of national policy and economic benefit to US foreign trade
  - Formulated U.S. policy of cooperating with friendly nations to develop nuclear power
    - *Provide an assured source of enriched uranium*
    - *Safeguards to prevent its diversion for military use*
  - Strongly supported centrifuge method for enriching uranium because of reduced power consumption compared with gaseous diffusion
    - *Differed with Commission's decision to delay development of centrifuge because of proliferation concerns*
    - *Argued that progress can't be stopped by legislation*
    - *Believed in using new developments and solving political problems instead*
- Resigned from AEC after only 4 years on February 1, 1964 due to failing health

Legacy

- **AEC Chairman Glenn T. Seaborg:**
  - "The entire atomic energy program will miss Dr. Wilson's services. He brought to the Commission not only an extensive technical background, but a broad experience in business and finance."
- **President Lyndon B. Johnson (in a letter to Robert E. Wilson):**
  - "Your outstanding performance as a commissioner and the high esteem and respect with which you are regarded by your fellow commissioners as a scientist, a businessman, and a public servant must be a source of great satisfaction to you as your years of public service come to an end. As a result of your foresight and determination, we have a stronger and more self-reliant private atomic energy industry today. I join all of your friends and a grateful nation in thanking you for your years of fruitful and beneficial service."
- **Robert E. Wilson died from a stroke 7 months later, on September 1, 1964**
  - While serving as an official adviser to the US delegation to the 3rd United Nations International Conference on the Peaceful Uses of Atomic Energy in Geneva, Switzerland.
- **Glenn T. Seaborg statement following Wilson's death:**
  - "Dr. Wilson's wide experience and wisdom, imparted with vigor and generous spirit, greatly enriched the development of atomic energy in the United States and in the world."

Acknowledgments

- **Primary sources of information:**
  - L. William Moore, Donald L. Campbell. "Robert Erastus Wilson, 1893-1964; A Biographical Memoir." National Academy of Sciences Biographical Memoirs, 1983. Available online at
    - http://www.nasonline.org/publications/biographical-memoirs/memoir-pdfs/wilson-robert.pdf
  - F.W. Woodfield. "The Robert E. Wilson Award – How It Happened." Paper 46a, Proceedings, AIChE Fall Meeting, Washington, DC (Diamond Jubilee), October 30-November 4, 1983.

598762 Analysis of Mixing Regimes for Refinery Hazardous Oily Sludge Biodegradation

Abstract:

Historic waste ponds containing oily sludge are commonly found at older oil refineries, and many of these ponds are not remediated or are still in minimal use at currently operating facilities. A common method for treating this sludge is by solidification / stabilization, if the waste unit is no longer in use. However, for currently existing waste ponds at operating refineries in which the units not under closure requirements or orders, biodegradation can be a viable and less-costly treatment option. In a quiescent wastewater pond environment, the oily sludge is present typically as a single phase. To promote aerobic biodegradation of the material, extensive mixing and aeration of the liquid will be necessary.

The research reported here investigated the level of mixing energy imparted to an oil phase present in a larger aqueous phase, in order to create oil droplets that would be amenable to biodegradation. The effectiveness of the mixing was measured using the dispersion effectiveness technique, which is an extraction approach using dichloromethane. The tests were conducted at both a laboratory scale (<0.5 liters) and a pilot scale (tens of liters), and with different types of mixing impellers. The experimental results indicated that higher mixing speeds led to a greater extent of oil dispersion, as is expected. However, there was a large drop in dispersion effectiveness with mixing speeds below approximately 150 rpm. Additionally, the energy dissipation was calculated with different mixing regimes (mixer speed and impeller type) and was compared with the dispersant effectiveness. This analysis indicated that similar levels of dispersant effectiveness were achievable using a helical impeller at low mixing speeds as compared with a marine impeller at higher mixing speeds. The success with the lower mixing speed of the helical impeller suggest that larger pond environments may be effectively mixed using less power with this type of mixing device.

593530 Antifoam Development for Processing of High Level Waste in the Defense Waste Processing Facility

Abstract:

Antifoam is needed during processing of High Level Waste (HLW) due to the high gas generation from steam and chemical reaction offgas products to prevent foam from contaminating the condensate. Antifoam is used to minimize foam[1] during chemical processing and HLW evaporation at the Savannah River Site (SRS)[2] and Hanford[3]. The antifoam used in the Defense Waste Processing Facility (DWPF) impacts flammability during chemical processing (generates three flammable degradation products) and while feeding the melter[4-5] since the antifoam can decompose to CO/hydrogen. It also is the likely source of methyl for the methyl mercury present in the tank farm and excessive mercury in Saltstone[6]. The planned startup of The Salt Waste Processing Facility (SWPF), with much higher throughput, will challenge the Defense Waste Processing Facility (DWPF) to process more efficiently.

DWPF employs Antifoam 747[7], a superspreader produced by Momentive Performance Materials, as an antifoaming agent during waste processing. Despite its ability to control foam, processing issues have risen from use, which has led to two "Potential Inadequacy in the Safety Analyses" (PISAs) and has limited DWPF facility throughput.

During DWPF chemical processing, antifoam must be effective up to 103°C between pH of 3-13. Antifoam 747 is most effective at pH 6-8[8] and degrades as pH deviates. Researchers at the Savannah River National Laboratory (SRNL) identified three flammable antifoam degradation[9] products using mass spec and Fourier Transformed InfraRed (FTIR) offgas analyzers during experiments[10].

A new Antifoam has been recommended for use in DWPF. The new antifoam was the best of the 30 antifoam candidates (identified by six antifoam manufacturers) that were evaluated in SRNL testing. The new antifoam is hydrolysis resistant which results that much less antifoam is needed for foam control. No degradation products were identified in the offgas or condensate. The testing leading to the new antifoam recommendation will be discussed.

1. Bindal, S. K.; Nikolov, A. D.; Wasan, D. T.; Lambert, D. P.; Koopman, D. C., Foaming in Simulated Radioactive Waste. *Sci. Technol.* **2001**,*35* (19), 3941-3947.
2. Wasan, D. T.; Lambert, D. P. *Foaming and Antifoaming in Radioactive Waste Pretreatment and Immobilization*; Illinois Institute of Technology: Chicago, IL, 2001.
3. Calloway, J. T. B.; Martino, C. J.; Jantzen, C. M.; Wilmarth, W. R.; Stone, M. E.; Pierce, R. A.; Josephs, J. E.; Barnes, C. D.; Daniel, W. E.; Eibling, R. E.; Choi, A. S.; White, T. L.; Crowley, D. A.; Baich, M. A.; Johnson, J. D.; Vijayaraghavan, K.; Nikolov, A. P.; Wasan, D. T., Radioactive Waste Evaporation: Current Methodologies Employed for the Development, Design and Operation of Waste Evaporators at the Savannah River Site and Hanford Waste Treatment Plant. **2003**,(37327), 157-170.
4. Choi, A. S.; Smith III, F. G.; McCabe, D. J. *Preliminary Analysis of Species Partitioning in the DWPF Melter. Sludge batch 7A*; SRNL-STI--2016-00540; Savannah River National Laboratory: Aiken, SC, 2017.
5. *Evaluation of the Safety of the Situation (ESS): Melter Feed Rate Temperature Correlation*

Basis (PISA PI-2014-0009); Savannah River Remediation LLC: Aiken, SC, 2016.

6. Meraw, H. J. *Volatilization and Flammability Characteristics of Elemental and Organic Mercury*; X-ESR-G-00048, Revision 2; Savannah River Remediation LLC: Aiken, SC, June 2015, 2015.
7. Koopman, D. C. *Comparison of Dow Corning 544 Antifoam to IIT747 Antifoam in the 1/240 SRAT*; WSRC-TR-99-00377; Savannah River Technology Center: Aiken, SC, 2000.
8. Lambert, D. P.; Koopman, D. C.; Newell, J. D.; Wasan, D. T.; Nikolov, A. P.; Weinheimer, E. K., Improved Antifoam Agent Study End of Year Report, EM Project 3.2.3. 2011.
9. McCord, J. B. *Evaluation of the Safety of the Situation (ESS): Melter Feed Rate Temperature Correlation Basis (PISA PI-2014-0009)*; Savannah River Remediation LLC: Aiken, SC, 2016.
10. Lambert, D. P.; Zamecnik, J. R.; Newell, J. D.; Martino, C. J. *Impact of Scaling on the Glycolic-Nitric Acid Flowsheet*; Savannah River National Laboratory: Aiken, SC, 2016.

603566 Decomposition of Nuclear Waste Spectra Using Blind Source Separation: Application to Hanford Waste

Abstract:

The clean-up efforts at the Hanford Site have been complicated by the nature of the waste, which includes many species and three different waste categories: supernatant liquid, saltcake solids and water-insoluble sludge [1]. Each of the 177 underground waste tanks has a unique combination of all the species and categories of waste, with lower variability for tanks in the same farm.

The supernatant liquid low-activity waste will be processed and stabilized via vitrification, as part of the Direct-Feed Low-Activity Waste (DFLAW) initiative. In our research, we use *in situ* infrared (attenuated total reflectance-Fourier transform infrared or ATR-FTIR) and Raman spectroscopy to measure the composition of the supernatant liquid, since *in situ* monitoring techniques can facilitate continuous operation of the DFLAW melter. In previous work [2], we applied Blind Source Separation (BSS) methods to quantify the composition of simpler mixtures containing six species.

In this contribution, we work with more complex mixtures, seeking to quantify the concentration of key species in the simulated waste. Examples of key nonradioactive species include nitrate and nitrite due to their prevalence in low-activity waste [3]. In order to obtain accurate estimates of the concentrations of these species, without performing lengthy calibration that span multiple levels of all the possible species, we developed a four-step procedure that couples BSS techniques with a standard calibration.

In the first two steps, we used BSS to: (1) identify all the independent components (i.e. species) in a mixture spectrum and (2) correlate the independent components with reference spectra from our library. At the end of step 2, we identified the contributions of the key species to the original mixture, which lets us proceed to: (3) correct the mixture spectrum by subtracting the additional information that pertains to the additional components. Now that we have the corrected mixture spectrum, we proceed to: (4) quantify the concentrations of the key species using a classical regression algorithm, such as Partial Least-Squares Regression (PLSR). The regression algorithm is trained with the species of interest only, thus reducing the training data set by hundreds of measurements.

The four-step procedure was applied to both simulated and experimental data, which include species of interest (nitrate, nitrite, carbonate and sulfate), as well as additional species (aluminate, phosphate, oxalate, chromate and acetate). Our research has shown that this method can be used to decompose the signal of complex mixtures to quantify specific components. The results are comparable to predictions from calibration approaches, which would have included lengthy experiments to collect spectra of all the possible combinations of major and minor components.

[1] Peterson, R. A.; Buck, E. C.; Chun, J.; Daniel, R. C.; Herting, D. L.; Ilton, E. S.; Lumetta, G. J.; Clark, S. B. Review of the Scientific Understanding of Radioactive Waste at the U.S. DOE Hanford Site. *Environmental Science & Technology* **2018**, 52, 108381–396

[2] Maggioni GM, Kocevska S, Grover MA, Rousseau RW. Analysis of Multicomponent Ionic Mixtures Using Blind Source Separation: A Processing Case Study. *Ind. Eng. Chem. Res.* **2019**, 58, 50, 22640-22651

[3] Howe AM. WTP Real-Time In-Line Monitoring Program Tasks 4 and 6: Data Quality and Management and Preliminary Analysis Plan. **2017**

611991 Ion-Exchange Modeling to Support the Tank Closure Cesium Removal Process at the Savannah River Site

Abstract:

The Tank Closure Cesium Removal (TCCR) system is a demonstration "at-tank" process designed to remove ^{137}Cs from the high-level aqueous tank waste at the Savanah River Site (SRS) so that the decontaminated solution can be disposed as low-level waste. Cesium is removed by ion exchange (IX) columns using engineered IONSIV[1] R9120-B of the Crystalline SilicoTitanate (CST) media.

The current TCCR system can accommodate lead-lag (two-column) or lead-lag-guard (three-column) configurations to optimize media utilization and achieve the target decontamination. To support the TCCR operation for decontamination of Tank 10 in the SRS H Tank Farm, Savannah River National Laboratory (SRNL) utilized the following modeling applications:

1. The ZAM computer program developed by Texas A&M University for calculation of equilibrium cesium loading on the CST media.
2. The VERSE-LC code developed by Purdue University to predict ion-exchange (IX) column performances.
3. The OLI Studio[2] commercial software from OLI Systems, Inc. to calculate charge-balanced waste compositions and to estimate waste solution properties (i.e., density, viscosity, diffusivity) required as input data to ZAM and VERSE-LC.

This paper demonstrates the versatility of the ion exchange modeling in evaluating the loading of cesium on CST and the plant operational IX column performances.

[1] IONSIV is a trademark of Honeywell UOP, Inc. Des Plaines, Illinois, USA

[2] OLI is a registered trademark of OLI Sytems, Inc., of Morris Plains, New Jersey, USA

Analysis of Radiolytic Hydrogen Generation from Water Adsorbed on Selected Plutonium Oxides

By: Dr. James E. Laurinat
Savannah River National Laboratory
Aiken, South Carolina

Abstract

Radiolytic hydrogen generation from adsorbed moisture poses a safety risk for storage of plutonium oxide powders in sealed containers. Measurements have shown that during long-term storage the hydrogen pressure in such containers may either reach an equilibrium value or go through a maximum and then slowly decrease. The tempering of the gas generation rate with increasing pressure can result either from competing gas generation and consumption reactions whose relative rates change with pressure or perhaps over time or from an equilibrium in gas pressure or dissolved gas activity. The existence of an equilibrium gas pressure or activity is a more attractive explanation from the standpoint that it does not require a second reaction mechanism that is separate from water radiolysis.

A model is proposed in which the rate of radiolytic gas generation is based on competing rates of generation and absorption of hydrogen and oxygen. To validate the model, a perturbation analysis is used to calculate a characteristic frequency for fluctuations in the pressure generated during a gas generation test. Calculated and measured frequencies are compared.

Introduction

In 1993, the U.S. Department of Energy (DOE) established the 3013 Standard for the Storage of Stabilized Plutonium Oxide and Plutonium Metal at DOE Facilities. The 3013 standard [1] establishes criteria for the long-term storage of these materials in sealed metal containers. A primary concern for 3013 containers is pressurization due to radiolytically generated hydrogen. Studies were conducted at the Los Alamos National Laboratory (LANL) and the Savannah River National Laboratory (SRNL) to measure the rate of pressurization. This paper develops a model to correlate the measured pressurization rate from a LANL study as a function of time, based on competing rates of radiolytic gas generation for hydrogen and oxygen. The relative rates of hydrogen and oxygen generation are modeled based on the results of hydrogen back pressure studies performed at SRNL. The model is extended to predict the frequency of long-term perturbations in measured backpressures. In the perturbation analysis, a simple gas-liquid diffusion model is applied. Thus, the perturbation model tests the hypothesis that the hydrogen gas pressure controls the rate of radiolytic gas generation in the liquid water adsorbed on the oxide.

Analysis of Radiolysis Stoichiometry from Hydrogen Back Pressure Tests

Duffey and Livingston measured rates of hydrogen absorption by plutonium oxide powders loaded with small amounts of moisture. [2] Their measurements, whose results are shown in Figure 1, demonstrated that the rate of pressure change decreases linearly as the hydrogen back pressure increases. The pressure change rate approaches zero at a partial hydrogen pressure of 25 psi for the reactor grade (Mark 42) oxide and a partial hydrogen pressure of 82 psi for the weapons grade (WG) oxide. The weapons grade and reactor grade oxides differ in their isotopic compositions and their firing (oxidation) temperatures. The weapons grade oxide is comprised of nearly pure Pu-239 and is considered low-fired, with an oxidation temperature of 450 °C. The reactor grade oxide contains Pu-239 with higher levels of other Pu isotopes and is considered high-fired, with an oxidation temperature of 700 °C. The fact that the rate of pressure change and therefore the rate of radiolytic gas generation are linear rather than logarithmic functions of the hydrogen pressure suggests that the hydrogen back pressure directly counteracts the activity, i.e., the pressure, exerted by the dissolved gases.

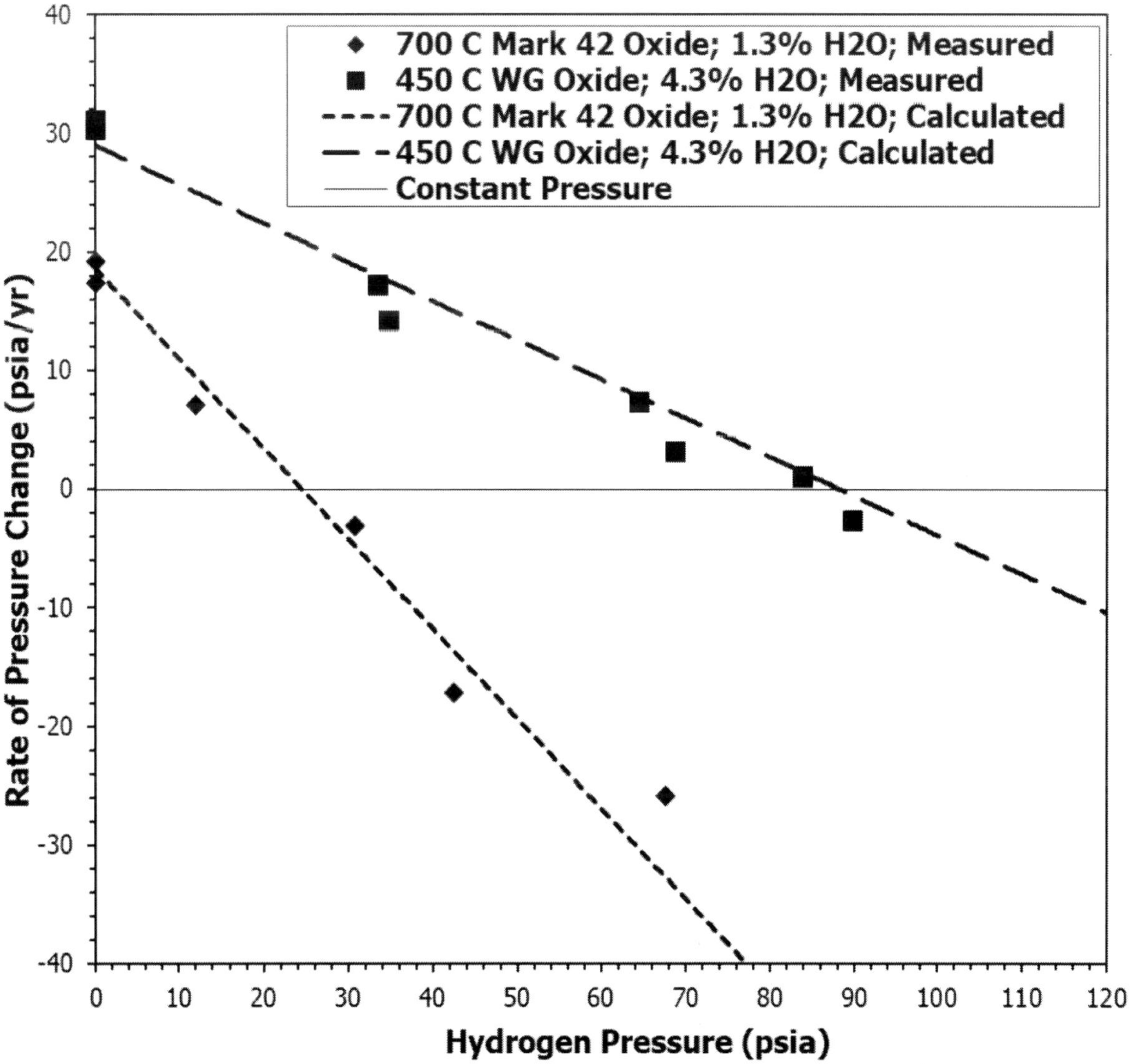

Figure 1. Comparison of Measured and Predicted Initial Rates of Pressure Change for Weapons Grade PuO_2 Oxidized at 450 °C and Reactor Grade PuO_2 Oxidized at 700 °C

The back pressure effect is somewhat similar to that observed during suppression of boiling. Nucleate boiling begins when a liquid is heated until its vapor pressure exceeds the ambient pressure. At that temperature, if the rate of heating exceeds the rate of heat loss due to evaporation, thermal convection, and thermal radiation, vapor bubbles form, and boiling commences. Similarly, radiolytic gas generation will occur provided that the partial pressure due to radiolysis exceeds the back pressure of the gas in contact with the oxide. The difference between boiling and radiolysis is that boiling is an inherently nonequilibrium process. Therefore, the vapor pressure must exceed the total pressure, including the partial pressures of the vapor and noncondensable gases, for boiling to take place. Radiolysis, by contrast, is a relatively slow process, so the vapor-liquid equilibrium is maintained. Consequently, radiolysis proceeds whenever the partial pressure of the radiolysis gases exceeds the partial pressure of the gas in contact with the oxide. For oxides stored at ambient or only slightly elevated temperatures so that the vapor pressure is low, this suggests that radiolysis will proceed until the equivalent partial pressure (or equivalent activity) of the radiolysis gases rises to the initial pressure for the storage container, which typically is one atmosphere at the local elevation.

In view of the preceding discussion, the approach to correlate the Duffey and Livingston data is to assume that some combination of the dissolved hydrogen and oxygen gases has an activity of one atmosphere. For the Duffey and Livingston hydrogen back pressure tests with the reactor grade oxide, it may be recalled that the controlling radiolysis reaction is the direct, stoichiometric water radiolysis reaction that generates two moles of hydrogen and one mole of oxygen for every two moles of water:

$$2H_2O(l) \xrightarrow{\alpha} 2H_2(l) + O_2(l) \qquad \text{(R-1)}$$

At the equilibrium pressure, the back pressure provided by the hydrogen must be equal to the combined pressures generated by the dissolved hydrogen and oxygen gases. In other words,

$$P_{H_2,b} - P_{H_2,s}^{*} = P_{H_2,s}^{*} + P_{O_2,s}^{*} \qquad (1)$$

where

$P_{H_2,s}^{*}$ = activity or partial pressure exerted by the dissolved hydrogen gas

$P_{O_2,s}^{*}$ = activity or partial pressure exerted by the dissolved oxygen gas

Solving for the hydrogen back pressure at equilibrium yields

$$P_{H_2,b} = 2P_{H_2,s}^{*} + P_{O_2,s}^{*} \qquad (2)$$

If it is assumed that the controlling step in the radiolysis reaction forms dissolved hydrogen and oxygen gases simultaneously, then the combined partial pressure exerted by the hydrogen and oxygen should sum to one atmosphere. Since water radiolysis generates two hydrogen molecules for every oxygen molecule, the individual partial pressures for hydrogen and oxygen at equilibrium can be assumed to be 2/3 and 1/3 atmosphere, respectively:

$$P_{H_2,s}^{*} = \frac{2}{3}P_a \qquad (3)$$

and

$$P_{O_2,s}^{*} = \frac{1}{3}P_a \qquad (4)$$

where

P_a = atmospheric pressure

Equations 2, 3, and 4 combine to give an equilibrium hydrogen back pressure of 5/3 atmospheres, or 24.5 psia:

$$P_{H_2,b} = \frac{5}{3} P_a \tag{5}$$

This equilibrium back pressure closely agrees with the value of 25 psia reported by Duffey and Livingston.

As discussed previously, radiolysis of the weapons grade oxide generates equimolar amounts of hydrogen and oxygen. In other words, if the radiolysis rate is based on the hydrogen generation rate, an extra oxygen molecule is generated for every oxygen molecule produced by radiolytic decomposition of a water molecule. Thus, at equilibrium the hydrogen back pressure is given by

$$P_{H_2,b} - P_{H_2,s}^{\ *} = P_{H_2,s}^{\ *} + 2P_{O_2,s}^{\ *} \tag{6}$$

or

$$P_{H_2,b} = 2P_{H_2,s}^{\ *} + 2P_{O_2,s}^{\ *} \tag{7}$$

It is apparent from the ratio of hydrogen to oxygen generation rates that the controlling step for the radiolysis of the weapons grade oxide is not simple water radiolysis. Instead, the mechanism that best explains the observed effect of the hydrogen back pressure is a mechanism whereby the limiting dissolved gas pressure for the radiolysis of water is the oxygen partial pressure. If this indeed is the limiting pressure, then the partial pressure exerted by the radiolysis is one atmosphere oxygen and two atmospheres hydrogen. Thus,

$$P_{O_2,s}^{\ *} = P_a \tag{8}$$

and

$$P_{H_2,s}^{\ *} = 2P_a \tag{9}$$

From equations 7, 8, and 9, then, the equilibrium hydrogen back pressure for radiolysis of the weapons grade oxide is six atmospheres, or 88.2 psia.

$$P_{H_2,b} = 6P_a \tag{10}$$

This pressure agrees reasonably well with the reported equilibrium pressure of 82 psia.

Pressure Transient Analysis

From the preceding discussion, it may be concluded that the partial pressure of hydrogen gas in excess of the stoichiometric ratio generated by radiolysis acts to suppress the generation of both hydrogen and oxygen, whereas any oxygen partial pressure in excess of that generated by radiolysis results in either the generation of additional hydrogen. It follows that the total gas pressure for water radiolysis approaches its ultimate equilibrium value when the hydrogen partial pressure in the gas reaches the total gas pressure generated by radiolysis. Thus, equilibrium for oxides similar to the Duffey and Livingston reactor grade oxide is achieved when

$$P_{H_2,b} = P_{H_2,s}^{\ *} + P_{O_2,s}^{\ *} = \frac{2}{3} P_a + \frac{1}{3} P_a = P_a \tag{11}$$

It may be assumed that the effective equilibrium constant for the total radiolysis pressure is given by the product of the hydrogen and oxygen pressures generated by radiolysis. Under this assumption, at equilibrium,

$$P_{O_2,b} = \frac{P_{H_2,s}^{\ *} P_{O_2,s}^{\ *}}{P_{H_2,s}^{\ *} + P_{O_2,s}^{\ *}} = \frac{\left(\frac{2}{3}P_a\right)\left(\frac{1}{3}P_a\right)}{\frac{2}{3}P_a + \frac{1}{3}P_a} = \frac{2}{9}P_a \qquad (12)$$

If one assumes that the gas space initially is occupied by an inert gas at one atmosphere, then

$$P_b = P_a + P_{O_2,b} + P_{H_2,b} = \frac{20}{9}P_a \qquad (13)$$

where

P_b = total bulk gas phase pressure

Pressure transients for plutonium oxide radiolysis can be estimated by solving rate equations for the gas phase diffusion limited rates of hydrogen and oxygen generation. ,This approach differs from that of other investigators, e.g. Pegg et al. [3] and Sims et al. [4] who model hydrogen generation rates using surface properties of the oxide powder.

To model the pressure transients, it is convenient to use a dimensionless time and dimensionless pressures that yield a dimensionless gas generation rate equal to one, for the generation of one molecule of hydrogen per radiolytic destruction of a water molecule. The dimensionless time is given by

$$\bar{t} = \frac{(0.001)(86400)\eta}{\Delta H_r}\frac{R_g T_i}{P_a V_g} R_\alpha x_{H_2O,free} t \qquad (14)$$

where

$\bar{t}$ = dimensionless time for radiolysis rate

R_g = gas law constant, 0.082057 L atm/mol/K

V_g = gas space volume, L

R_α = radiolytic heat generation rate, w

t = time, day

η = efficiency of radiolysis reaction

$x_{H_2O,free}$ = fraction of total electrons in the oxide that belong to free (unbound water)

The pressures are nondimensionalized by dividing the partial pressures of hydrogen and oxygen by the atmospheric pressure, to obtain the dimensionless pressures $\bar{P}_{H_2,b}$ and $\bar{P}_{O_2,b}$, where

$\bar{P}_{H_2,b}$ = dimensionless bulk hydrogen pressure in the gas, normalized with respect to the initial atmospheric pressure

$\bar{P}_{O_2,b}$ = dimensionless bulk oxygen pressure in the gas, normalized with respect to the initial atmospheric pressure

An analysis of the pressure transient for a typical oxide stored at LANL, MIS Item No. 11589A, follows. This item contains plutonium oxide and salts which apparently are inert to radiolytic attack, so the material in this item will be referred to as a 'pure' oxide.

Because the exact forms of the rate equations for hydrogen and oxygen generation are not known, a linearizing approximation is made in which all effects of the hydrogen and oxygen back pressures on the generation rates are modeled using linear coefficients. The linearized generation rate equations for hydrogen and oxygen generation take the forms

$$\frac{d\bar{P}_{H_2,b}}{d\bar{t}} = C_H - C_{H-H}\bar{P}_{H_2,b} - C_{H-O}\bar{P}_{O_2,b} \tag{15}$$

and

$$\frac{d\bar{P}_{O_2,b}}{d\bar{t}} = C_O - C_{O-H}\bar{P}_{H_2,b} - C_{O\ O}\bar{P}_{O_2,b} \tag{16}$$

where

C_H = constant representing rate of change of hydrogen pressure with no back pressure

C_{H-H} = coefficient for the effect of hydrogen back pressure on the rate of change of the hydrogen pressure

C_{H-O} = coefficient for the effect of oxygen back pressure on the rate of change of the hydrogen pressure

C_O = constant representing rate of change of oxygen pressure with no back pressure

C_{O-O} = coefficient for the effect of oxygen back pressure on the rate of change of the oxygen pressure

C_{O-H} = coefficient for the effect of hydrogen back pressure on the rate of change of the oxygen pressure

The hydrogen and oxygen rate equations are separated by making each rate equation second order, and back substituting to eliminate nonhomogeneous terms. Accordingly, a time derivative of the hydrogen generation rate equation gives

$$\frac{d^2\bar{P}_{H_2,b}}{d\bar{t}^2} = -C_{H-H}\frac{d\bar{P}_{H_2,b}}{d\bar{t}} - C_{H-O}\frac{d\bar{P}_{O_2,b}}{d\bar{t}} \tag{17}$$

Substitution for $\frac{d\bar{P}_{O_2,b}}{d\bar{t}}$ by equation 16 then yields

$$\frac{d^2\bar{P}_{H_2,b}}{d\bar{t}^2} = -C_{H-H}\frac{d\bar{P}_{H_2,b}}{d\bar{t}} - C_O C_{H-O} + C_{H-O}C_{O-H}\bar{P}_{H_2,b} + C_{H-O}C_{O-O}\bar{P}_{O_2,b} \tag{18}$$

and substitution for $\bar{P}_{O_2,b}$ from equation 15 gives

$$\frac{d^2\bar{P}_{H_2,b}}{d\bar{t}^2} = -\left(C_{H-H} + C_{O-O}\right)\frac{d\bar{P}_{H_2,b}}{d\bar{t}} + \left(C_{H-O}C_{O-H} - C_{H-H}C_{O-O}\right)\bar{P}_{H_2,b} + C_H C_{O-O} - C_O C_{H-O} \tag{19}$$

An analogous process for the oxygen rate equation results in the following equations:

$$\frac{d^2\bar{P}_{O_2,b}}{d\bar{t}^2} = -C_{O-H}\frac{d\bar{P}_{H_2,b}}{d\bar{t}} - C_{O-O}\frac{d\bar{P}_{O_2,b}}{d\bar{t}} \tag{20}$$

$$\frac{d^2\bar{P}_{O_2,b}}{d\bar{t}^2} = -C_H C_{O-H} + C_{H-H}C_{O-H}\bar{P}_{H_2,b} + C_{H-O}C_{O-H}\bar{P}_{O_2,b} - C_{O-O}\frac{d\bar{P}_{O_2,b}}{d\bar{t}} \tag{21}$$

and

$$\frac{d^2\bar{P}_{O_2,b}}{d\bar{t}^2} = -\left(C_{H-H} + C_{O-O}\right)\frac{d\bar{P}_{O_2,b}}{d\bar{t}} + \left(C_{H-O}C_{O-H} - C_{H-H}C_{O-O}\right)\bar{P}_{O_2,b} + C_O C_{H-H} - C_H C_{O-H} \tag{22}$$

As $\bar{t} \to \infty$,

$$\frac{d\bar{P}_{H_2,b}}{d\bar{t}} = 0 \tag{23}$$

$$\frac{d^2\bar{P}_{H_2,b}}{d\bar{t}^2} = 0 \tag{24}$$

$$\frac{d\bar{P}_{O_2,b}}{d\bar{t}} = 0 \tag{25}$$

and

$$\frac{d^2\bar{P}_{O_2,b}}{d\bar{t}^2} = 0 \tag{26}$$

Therefore, there are no particular solutions (i.e., solutions containing polynomials of $\bar{t}$) of equations 19 and 22. In other words, the total solutions to these differential equations consist of the general solution and a constant that represents the equilibrium pressure.

It may be noted that, except for the constant terms, equations 19 and 22 are identical. The time derivative expressions for both equations may be factored by solving the quadratic expression

$$D^2 + \left(C_{H-H} + C_{O-O}\right)D + C_{H-H}C_{O-O} - C_{H-O}C_{O-H} = 0 \tag{27}$$

to obtain

$$D = \frac{-C_{H-H} - C_{O-O} \pm \sqrt{\left(C_{H-H} + C_{O-O}\right)^2 - 4C_{H-H}C_{O-O} + 4C_{H-O}C_{O-H}}}{2} \tag{28}$$

or

$$D = \frac{-C_{H-H} - C_{O-O} \pm \sqrt{\left(C_{H-H} - C_{O-O}\right)^2 + 4C_{H-O}C_{O-H}}}{2} \tag{29}$$

Equation 29 can be simplified by utilizing relationships among the back pressure coefficients C_{H-H}, C_{O-O}, C_{O-H}, and C_{H-O}. The effects of the hydrogen back pressure on the hydrogen generation rate and of the oxygen back pressure on the oxygen generation rate are identical. Thus,

$$C_{H-H} = C_{O-O} \tag{30}$$

For plutonium oxides similar to the Duffey and Livingston reactor grade material, with equilibrium pressures given by equations 11 through 13, the generation of two hydrogen molecules from the radiolysis of one water molecule is accompanied by the stoichiometric generation of one oxygen molecule. As noted previously, results from the Duffey and Livingston tests indicate that the hydrogen back pressure controls the overall radiolysis rate, so the hydrogen back pressure may also be viewed as controlling the oxygen generation rate by radiolysis. Therefore,

$$C_{O-H} = \frac{1}{2} C_{H-H} \tag{31}$$

The fact that the hydrogen back pressure limits the radiolysis rate implies that the excess oxygen pressure in the gas space affects the gas generation rate by oxidizing the plutonium oxide and thereby releasing additional hydrogen. Consequently, the sign of C_{H-O} is opposite to the sign of C_{H-H}. It is assumed that the oxidation reaction follows the stoichiometry of Reaction R-1, in which a half mole of oxygen is consumed and two moles of hydrogen are generated:

$$6PuOOH(s) + O_2(l) + H_2O(l) \xrightarrow{(\alpha)} 6PuO_2 \bullet \frac{1}{3} H_2O(s) + 2H_2(l) + \frac{1}{2} O_2(l) \tag{R-2}$$

The oxidation reaction represents conversion of excess oxygen generated by the stoichiometric radiolysis reaction, whose rate, one may recall, is controlled by the hydrogen back pressure. It may be concluded that the coefficient C_{H-O} is linked to the coefficient C_{O-H}, with a stoichiometry factor equal to four. In other words,

$$C_{H-O} = -4C_{O-H} = -2C_{H-H} \tag{32}$$

Substitution of equations 30, 31, and 32 in equation 29 yields

$$D = -C_{H-H} \pm iC_{H-H} \tag{33}$$

From equation 33, the real components of the complex general solutions for the hydrogen and oxygen pressures take the form

$$\bar{P}_{H_2,b} = \left(A_H \cos\left(C_{H-H}\bar{t}\right) + B_H \sin\left(C_{H-H}\bar{t}\right)\right)\exp\left(-C_{H-H}\bar{t}\right) + C_H \tag{34}$$

and

$$\bar{P}_{O_2,b} = \left(A_O \cos\left(C_{H-H}\bar{t}\right) + B_O \sin\left(C_{H-H}\bar{t}\right)\right)\exp\left(-C_{H-H}\bar{t}\right) + C_O \tag{35}$$

where

A_H = coefficient in general solution for hydrogen partial pressure

B_H = coefficient in general solution for hydrogen partial pressure

A_O = coefficient in general solution for oxygen partial pressure

B_O = coefficient in general solution for oxygen partial pressure

The coefficients A_H, B_H, A_O, and B_O and the constants C_H and C_O are determined from initial and final conditions. As $\bar{t} \to \infty$,

$$\bar{P}_{H_2,b} = 1 \tag{36}$$

and

$$\bar{P}_{O_2,b} = \frac{2}{9} \tag{37}$$

Consequently,

$$C_H = 1 \tag{38}$$

and

$$C_O = \frac{2}{9} \tag{39}$$

When $\bar{t} = 0$,

$$\bar{P}_{H_2,b} = 0 \tag{40}$$

and

$$\bar{P}_{O_2,b} = \bar{P}_{O_2,b,i} \tag{41}$$

Thus,

$$A_H = -1 \tag{42}$$

and

$$A_O = \bar{P}_{O_2,b,i} - \frac{2}{9} \tag{43}$$

Additionally, when $\bar{t} = 0$,

$$\frac{d\bar{P}_{H_2,b}}{d\bar{t}} = 1 \tag{44}$$

and

$$\frac{d\bar{P}_{O_2,b}}{d\bar{t}} = \frac{1}{2} \tag{45}$$

Therefore,

$$B_H = \frac{1 - C_{H-H}}{C_{H-H}} \tag{46}$$

and

$$B_O = \frac{\frac{1}{2} + \left(\bar{P}_{O_2,b,i} - \frac{2}{9}\right) C_{H-H}}{C_{H-H}} \tag{47}$$

With the application of the initial and final conditions, the solutions for the hydrogen and oxygen pressures become

$$\bar{P}_{H_2,b} = \left(-\cos\left(C_{H-H}\bar{t}\right) + \frac{1 - C_{H-H}}{C_{H-H}} \sin\left(C_{H-H}\bar{t}\right) \right) \exp\left(-C_{H-H}\bar{t}\right) + 1 \tag{48}$$

and

$$P_{O_2,b} = \left(\left(\bar{P}_{O_2,b,i} - \frac{2}{9} \right) \cos\left(C_{H-H}\bar{t}\right) + \frac{\frac{1}{2} + \left(\bar{P}_{O_2,b,i} - \frac{2}{9} \right) C_{H-H}}{C_{H-H}} \sin\left(C_{H-H}\bar{t}\right) \right) \exp\left(-C_{H-H}\bar{t}\right) + \frac{2}{9} \qquad (49)$$

It may be noted that these solutions contain the undetermined coefficient C_{H-H}. This coefficient is calculated by requiring that the pressures vary at a natural frequency at which gains in the real component of the pressure variations are offset by losses in the imaginary component, and vice versa. This requirement is satisfied if the integral of the real component over the entire time range equals zero. Different values of C_{H-H} would be obtained by applying this requirement to the hydrogen and oxygen pressures. Because it has been determined that the hydrogen back pressure controls the overall gas generation rate, the requirement is applied to the hydrogen pressure. Thus, the following condition is applied to calculate C_{H-H}:

$$\int_0^\infty \left(\bar{P}_{H_2,b} - 1 \right) d\bar{t} = 0 \qquad (50)$$

or

$$\int_0^\infty \left(-\cos\left(C_{H-H}\bar{t}\right) + \frac{1 - C_{H-H}}{C_{H-H}} \sin\left(C_{H-H}\bar{t}\right) \right) \exp\left(-C_{H-H}\bar{t}\right) d\bar{t} = 0 \qquad (51)$$

Integration by parts gives

$$\int_0^\infty \left(\frac{1 - C_{H-H}}{C_{H-H}} - 1 \right) \cos\left(C_{H-H}\bar{t}\right) \exp\left(-C_{H-H}\bar{t}\right) d\bar{t} = 0 \qquad (52)$$

This integral expression is satisfied if

$$\frac{1 - C_{H-H}}{C_{H-H}} - 1 = 0 \qquad (53)$$

or

$$C_{H-H} = \frac{1}{2} \qquad (54)$$

Substitution of equation 54 in equations 48 and 49 gives the final expressions for the hydrogen and oxygen pressures:

$$\bar{P}_{H_2,b} = \left(-\cos\left(\frac{\bar{t}}{2}\right) + \sin\left(\frac{\bar{t}}{2}\right) \right) \exp\left(-\frac{\bar{t}}{2}\right) + 1 \qquad (55)$$

and

$$\bar{P}_{O_2,b} = \left(\left(\bar{P}_{O_2,b,i} - \frac{2}{9} \right) \cos\left(\frac{\bar{t}}{2}\right) + \left(\bar{P}_{O_2,b,i} + \frac{7}{9} \right) \sin\left(\frac{\bar{t}}{2}\right) \right) \exp\left(-\frac{\bar{t}}{2}\right) + \frac{2}{9} \qquad (56)$$

The total pressure is computed by summing the hydrogen and oxygen pressures and the initial pressure, assumed to be one atmosphere. This gives

$$\bar{P}_{total} = \left(\left(\bar{P}_{O_2,b,i} - \frac{11}{9} \right) \cos\left(\frac{\bar{t}}{2}\right) + \left(\bar{P}_{O_2,b,i} + \frac{16}{9} \right) \sin\left(\frac{\bar{t}}{2}\right) \right) \exp\left(-\frac{\bar{t}}{2}\right) + \frac{20}{9} \qquad (57)$$

where

$\bar{P}_{total}$ = total pressure due to radiolysis, normalized with respect to the initial atmospheric pressure

The maximum total pressure may now be calculated by setting the time derivative of the total pressure equal to zero:

$$\left.\frac{d\bar{P}_{total}}{d\bar{t}}\right|_{\bar{P}_{total,max},\,\bar{t}_{max}} = 0 \tag{58}$$

This gives

$$\bar{P}_{total,max} = \frac{\sqrt{754}}{18}\exp\left(-\frac{1}{2}\left(\pi - \sin^{-1}\left(\frac{135}{377}\right)\right)\right) + \frac{20}{9} = 2.603 \tag{59}$$

$$\text{at } \bar{t}_{max} = \pi - \sin^{-1}\left(\frac{12}{37}\right) = 2.775 \tag{60}$$

where

$\bar{P}_{total,max}$ = maximum total pressure due to radiolysis

$\bar{t}_{max}$ = time at which the maximum total pressure due to radiolysis is reached

Figure 2 compares the measured transients for MIS Item No. 11589 [5] with the predictions of the linearized and corrected equilibrium-based gas generation rate models. The predicted curves were fit to the measured time scale using a single measured pressure, the hydrogen partial pressure at 306 days. The linear model fits the initial portions of the measured pressure transients well and accurately predicts both the magnitude and relative timing of the maximum hydrogen and oxygen partial pressures.

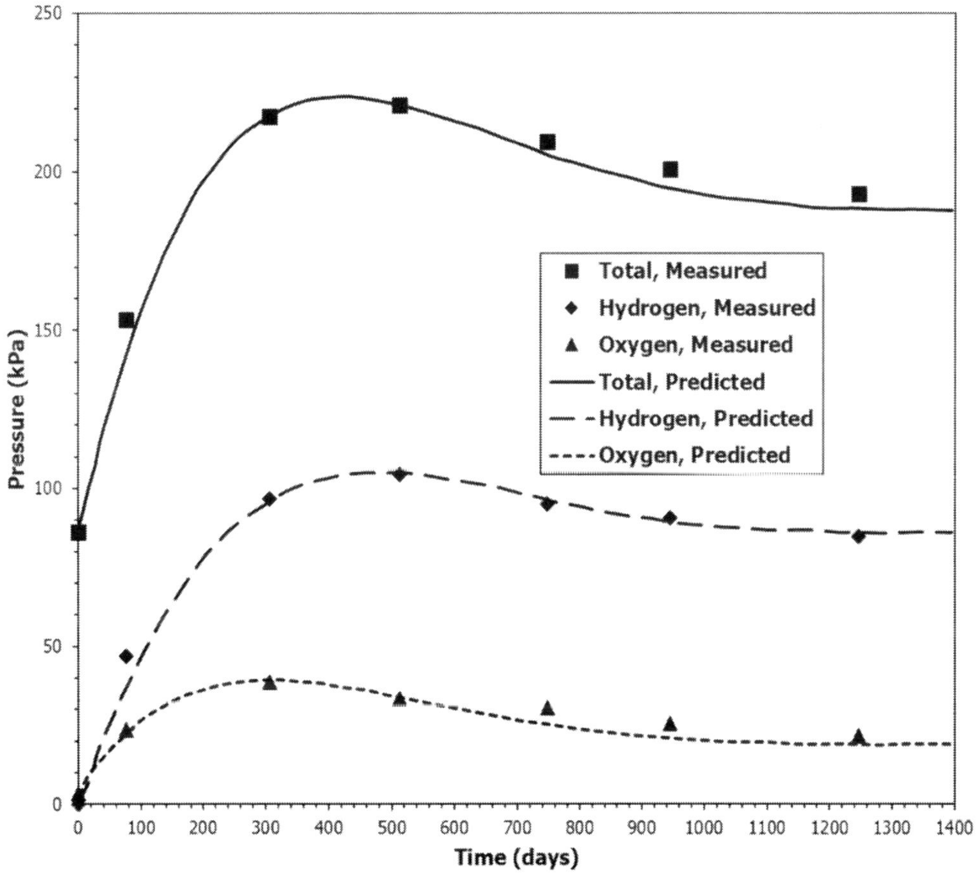

Figure 2. Comparison of Measured and Predicted Pressures for Radiolytic Gas Generation for MIS Item 11589A.

It may be noted that the total pressure shown in this figure includes the partial pressures of inert helium and nitrogen initial present in the container and therefore exceeds the sum of the hydrogen and oxygen partial pressures.

Perturbation Analysis of Pressure Transients for Storage of Plutonium Oxide Samples

A typical pressure transient has been examined to search for pressure oscillations. This transient was for gas generation from weapons grade oxide with 4.3 wt % moisture with 3450 torr initial hydrogen back pressure. (This test represents the 65 psi datum point in Figure 1. The transient pressures at this equilibrium condition were obtained from the authors of Reference 2. [6]) This transient test was accompanied by another test using a blank with 3000 torr initial hydrogen pressure. In both tests, the vessel was repeatedly purged with hydrogen prior to the test, so that the gas space inside the vessel initially was pure hydrogen except for approximately 20 torr of water vapor. Simultaneous temperature transients were recorded so that the pressure transients could be corrected for the compressibility of the gas in the vessel due to temperature fluctuations. The examination of the two pressure transients was limited to the temperature-corrected pressure signals.

To examine the pressure oscillations, a least squares regression was performed to calculate the average linear rate of pressure increase. This linear rate was then subtracted from the measured transient to capture just the fluctuating component of the pressure signals. Finally, the fluctuating components were analyzed for any characteristic periodic oscillations using a Lomb periodogram, which calculates dominant frequencies for signals over irregularly spaced intervals. [7, 8]

A linear perturbation analysis can be used to estimate the dominant frequency for a pressure perturbation due to radiolytic gas generation. The general approach of linear perturbation analyses is to formulate a rate equation for a perturbation, assume that there is a harmonic (sinusoidal) perturbation, and then to calculate the fastest growing and therefore dominant frequency for that perturbation by means of a complex number calculation.

The premise for the analysis of the pressure transient oscillations is to assume that there is a perturbation in the rate of gas generation at the gas-liquid interface. Prior to this perturbation, it is assumed that the interface is in equilibrium in that the rate of transport of dissolved hydrogen and oxygen from the liquid to the surface equals the rate of generation of hydrogen and oxygen gases. It follows from this assumption that half of the perturbation must propagate through the liquid phase and the other half must propagate through the gas phase. The following analysis is restricted to the half of the perturbation that propagates to the liquid phase. Since the test being analyzed had only hydrogen gas, all dissolved gas must be hydrogen. However, the analysis is for pressure perturbations, so all generation of both hydrogen and oxygen gases is included. Consequently, the mass balance for the perturbation reduces to

$$\frac{dc_{H_2}}{dt} = k_{gen} \qquad (61)$$

where

c_{H_2} = concentration of dissolved hydrogen gas

k_{gen} = rate of radiolytic generation of dissolved gas per unit volume of 'free' water

t = time

It may be noted that the radiolytic generation rate is based on the volume of 'free' water. This follows from the assumption that the portion of the water that is bound to the oxide is not able to absorb dissolved gases.

Following the standard procedure for a linear perturbation analysis, the time derivative in the mass balance is expressed as a wave number α

$$\alpha c_{H_2} = k_{gen} \qquad (62)$$

where

α = angular frequency for oscillations in dissolved hydrogen gas concentration

All variables are now written out as complex numbers, so that

$$(\alpha_R + i\alpha_I)(c_{H_2,R} + ic_{H_2,I}) = k_{gen,R} + i\,k_{gen,I} \qquad (63)$$

where

$i = (-1)^{0.5}$

$c_{H_2,I}$ = imaginary component of dissolved hydrogen gas concentration

$c_{H_2,R}$ = real component of dissolved hydrogen gas concentration

$k_{gen,I}$ = imaginary component of the rate of gas generation per unit volume of 'free' water

$k_{gen,R}$ = real component of the rate of gas generation per unit volume of 'free' water

α_I = imaginary component of angular frequency for oscillations in dissolved hydrogen gas concentration

α_R = real component of angular frequency for oscillations in dissolved hydrogen gas concentration

The real and imaginary components of equation 63 can be separated and solved. The real component is

$$\alpha_R c_{H_2,R} - \alpha_I c_{H_2,I} = k_{gen,R} \tag{64}$$

and the imaginary component is

$$\alpha_R c_{H_2,I} + \alpha_I c_{H_2,R} = k_{gen,I} \tag{65}$$

These equations can be solved for the real and imaginary components of the wave number to get

$$\alpha_R = \frac{k_{gen,R} c_{H_2,R} + k_{gen,I} c_{H_2,I}}{c_{H_2,R}^{\ 2} + c_{H_2,I}^{\ 2}} \tag{66}$$

and

$$\alpha_I = \frac{k_{gen,I} c_{H_2,R} - k_{gen,R} c_{H_2,I}}{c_{H_2,R}^{\ 2} + c_{H_2,I}^{\ 2}} \tag{67}$$

Subsequent steps in the analysis are simplified if the imaginary components of the generation rate and dissolved gas concentration are normalized with respect to the real components. This yields

$$\alpha_R = \frac{k_{gen,R}}{c_{H_2,R}} \frac{1 + a_k a_c}{1 + a_c^{\ 2}} \tag{68}$$

and

$$\alpha_I = \frac{k_{gen,R}}{c_{H_2,R}} \frac{a_k - a_c}{1 + a_c^{\ 2}} \tag{69}$$

where

$$a_k = \frac{k_{gen,I}}{k_{gen,R}} \tag{70}$$

and

$$a_c = \frac{c_{H_2,I}}{c_{H_2,R}} \qquad (71)$$

The terms a_k and a_c represent the tangents of the phase angles for the generation rate and the dissolved gas concentration, respectively.

From the premise that the perturbation is driven by the difference between the local rate of production of dissolved gases and the thermodynamically limited rate of gas generation, a_k is set equal to ratio of the excess in the production rate to the gas generation rate. Because the real and imaginary components of the generation rate are, by definition, orthogonal to each other, they combine as a root sum of squares. Thus,

$$a_k = \frac{\left(\eta_{max}{}^2 - \eta^2\right)^{0.5}}{\eta} \qquad (72)$$

where

η_{max} = maximum theoretical efficiency for the radiolysis reaction

The maximum efficiency is computed from the assumption that the amount of energy supplied for radiolysis is sufficient to overcome any energy barriers for intermediate steps in the overall radiolysis reaction mechanism, so that the limiting energy requirement is that for the alpha-catalyzed radiolysis of liquid water. This assumption likely is valid if the thermodynamic efficiency is low, as will be argued, so that there is much more energy available than can be utilized by the overall reaction.

Under this assumption, the G-value for hydrogen generation is just the inverse of the heat of reaction, converted from a molar to a molecular basis, multiplied by the thermodynamic efficiency. Thus,

$$G(H_2) = \frac{\left(1.602 \times 10^{-20}\right)\eta N_A}{\Delta H_r} \qquad (73)$$

where

$G(H_2)$ = intrinsic G-value for hydrogen generation based on the amount of water present, molec/100 eV

N_A = Avogadro's number, 6.022 x 10^{23} molec/mol

ΔH_r = heat of reaction for the radiolysis reaction, kJ/mol

The heat of reaction is calculated from the heats of formation for liquid water, hydrogen, and oxygen:

$$\Delta H_r = \Delta H_{f,H_2} + 0.5\Delta H_{f,O_2} - \Delta H_{f,H_2O,l} \qquad (74)$$

where

$\Delta H_{f,H_2}$ = heat of formation for hydrogen gas, kJ/mol

$\Delta H_{f,O_2}$ = heat of formation for oxygen gas, kJ/mol

$\Delta H_{f,H_2O,l}$ = heat of formation for liquid water, kJ/mol

The heats of formation for hydrogen and oxygen are zero by definition, so the heat of reaction is equal to the negative of the heat of formation for water, which is 285.8 kJ/mol.[1]

The efficiency of the radiolytic gas generation reaction is limited by the fact that the energy transfer from the alpha particles to water molecules is not 100% efficient. The maximum efficiency for alpha radiolysis, reported in terms of the G-value for the disappearance of water molecules, $G(H_2O)$, is reported to be 2.71 molec/100 eV.[2] This yields a computed maximum efficiency, η_{max}, of 0.0803, based on the heat of reaction for the hydrolysis reaction.

The actual efficiency η is derived thermodynamically. The radiolysis process can be considered to be a reversible Carnot cycle, for which the efficiency is given by the temperature ratio:

$$\eta = \frac{T_i - T_f}{T_i} \tag{75}$$

where

T_i = local initial temperature before the radiolysis reaction, K

T_f = local final temperature after the radiolysis reaction, K

The Arrhenius expression relating the surface vapor pressure during gas evolution to the bulk gas vapor pressure takes the form:

$$\frac{P_{v,H_2O,s}}{P_{v,H_2O,b}} = \exp\left(-\frac{E_a}{R_g}\left(\frac{1}{T_f} - \frac{1}{T_i}\right)\right) \tag{76}$$

where

E_a = activation energy in Arrhenius expression for water vapor pressure, J/mol

R_g = gas law constant, 8.314 J/mol/K

The Arrhenius constant for the water vapor pressure, $\frac{E_a}{R_g}$, is 5238.4 K.

Equations 3 and 20 can be rearranged to give

$$T_f = \frac{T_i}{1 + \frac{R_g T_i}{E_a} \ln\left(\frac{P_{v,H_2O,b}}{P_{v,H_2O,s}}\right)} \tag{77}$$

and

$$\eta = \frac{\frac{R_g T_i}{E_a} \ln\left(\frac{P_{v,H_2O,b}}{P_{v,H_2O,s}}\right)}{1 + \left|\frac{R_g T_i}{E_a} \ln\left(\frac{P_{v,H_2O,b}}{P_{v,H_2O,s}}\right)\right|} \tag{78}$$

Note that the vapor pressure term in the denominator of equation 78 is defined as an absolute value. This definition accommodates cases where radiolysis results in recombination of hydrogen and oxygen to form water, accompanied by a pressure decrease.

It can be shown that, for gas mixtures that are dilute in hydrogen, oxygen, and water vapor, the surface temperature at the reaction site drops so that the vapor pressure after the reaction occurs is 2/5 of the vapor pressure prior to the reaction. (If the surface water vapor pressure is 2/5 of the water vapor pressure in the bulk gas far from the surface, the partial pressure of the hydrogen and oxygen generated by the radiolysis reaction is 3/5, and the total partial pressure of water vapor and radiolysis gases at the surface is equal to the water vapor pressure in the bulk gas.) Thus,

$$P_{v,H_2O,s} \cong 0.4 P_{v,H_2O,b} \tag{79}$$

where

$P_{v,H_2O,b}$ = initial water vapor pressure in bulk gas

$P_{v,H_2O,s}$ = water vapor pressure at the surface after the radiolysis reaction

The value for a_c is calculated by maximizing the rate of growth of the perturbation, which is represented by the real component of the wave number:

$$\frac{\partial \alpha_R}{\partial a_c} = 0 \tag{80}$$

Solution of equations 68 and 80 gives

$$a_k = \frac{2a_c}{1 - a_c^{\,2}} \tag{81}$$

Substitution of equation 81 in equations 69 and 70 yields the solutions for the wave number components:

$$\alpha_I = \frac{k_{gen,R}}{c_{H_2,R}} \frac{a_k}{2} \tag{82}$$

and

$$\alpha_R = \frac{k_{gen,R}}{c_{H_2,R}} \frac{\left(1 + a_k^{\,2}\right)^{0.5} + 1}{2} \tag{83}$$

In terms of the maximum and thermodynamic radiolysis efficiencies, these solutions take the forms

$$\alpha_I = \frac{k_{gen,R}}{c_{H_2,R}} \frac{\left(\eta_{max}^{\,2} - \eta^2\right)^{0.5}}{2\eta} \tag{84}$$

and

$$\alpha_R = \frac{k_{gen,R}}{c_{H_2,R}} \frac{\eta_{max} + \eta}{2\eta} \tag{85}$$

The dominant frequency and period for the perturbation now can be computed given values for the real components of the generation rate and the dissolved gas concentration. It is assumed that dissolved gas is well-mixed, so that the ratio of the perturbation values for the generation rate and the dissolved gas concentration equals the ratio of the equilibrium values for these two quantities. The liquid is saturated with dissolved hydrogen gas at equilibrium. Thus,

$$\frac{k_{gen,R}}{c_{H_2,R}} = \frac{K_{gen}}{c_{H_2,sat}} \tag{86}$$

where

$c_{H_2,sat}$ = dissolved hydrogen gas concentration at saturation

K_{gen} = equilibrium rate of gas generation

At saturation, the dissolved hydrogen gas concentration equals the hydrogen partial pressure in the gas phase divided by the Henry's law constant for hydrogen, which is 1282.05 atm/M.

$$c_{H_2,sat} = \frac{P_{H_2}}{H_{H_2}} \tag{87}$$

where

H_{H_2} = Henry's law constant for hydrogen gas dissolved in water

P_{H_2} = partial pressure of hydrogen gas

The equilibrium gas generation rate is given by the rate of molar gas generation, computed from the rate of pressure increase, per unit volume of 'free' water, where the rate of pressure increase is converted into the gas generation rate via the ideal gas law:

$$K_{gen} = \frac{V_g}{2R_g T V_{H_2O}} \frac{f_{H_2O}}{f_{H_2O,free}} \frac{dP}{dt} \tag{88}$$

where

V_{H_2O} = volume of water

V_g = gas space volume

$\frac{dP}{dt}$ = rate of pressure increase calculated by linear regression of data

f_{H_2O} = weight fraction of water on oxide

$f_{H_2O,free}$ = weight fraction of water that is not bound to the oxide surface

The factor two appears in equation 81 because only half of the perturbed gas generation rate propagates into the liquid.

By definition, the imaginary component of the wave number is related to the frequency by

$$\alpha_I = 2\pi f \tag{89}$$

where

f = temporal frequency of oscillations in dissolved hydrogen gas concentration

Substitution of equations 86 through 89 in equation 84 gives the final expression for the frequency:

$$f = \frac{H_{H_2} V_g}{8\pi P_{H_2} R_g T V_{H_2O}} \frac{f_{H_2O}}{f_{H_2O,free}} \frac{dP}{dt} \frac{\left(\eta_{max}{}^2 - \eta^2\right)^{0.5}}{\eta} \tag{90}$$

The dominant period, τ, is just the reciprocal of the frequency.

$$\tau = \frac{1}{f} \tag{91}$$

From the preceding analysis, the dominant period for the propagation of perturbations in the gas generation rate into the liquid phase is estimated to be 268 hr. This is quite close to the dominant period of 254 hr identified by the Lomb periodogram, as Figure 3 shows. One may conclude that the linear perturbation analysis demonstrates that the observed fluctuations in the pressure signal from this test may indeed result from an overdriven gas generation rate, as predicted by the model.

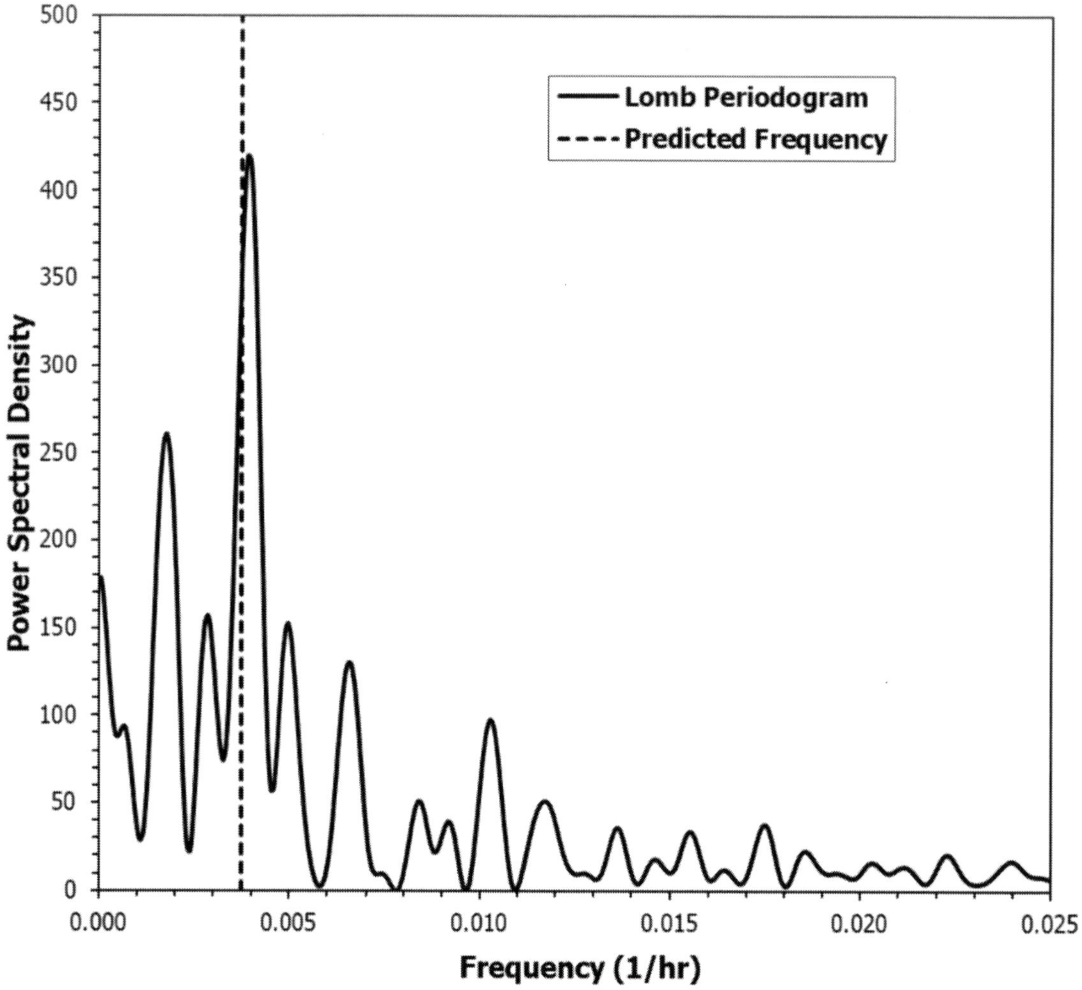

Figure 3. Lomb Periodogram for Pressure Deviation from Linear Rate of Increase during Test of Weapons Grade Material with 4.3 wt % Moisture and 3450 torr Initial Hydrogen Pressure

Conclusions

Analytic rate expressions have been derived for a plutonium oxide material containing weapons grade plutonium. The rate expressions are derived by linearizing the effects of hydrogen and oxygen back pressures in the gas space on the hydrogen and oxygen generation rates. The derived rate expressions accurately correlate the pressure transient measured by Los Alamos National Laboratory for storage of a typical plutonium oxide powder in a sealed test container.

Transient radiolytic gas generation rates fluctuate with a characteristic frequency. A linear perturbation analysis shows that this frequency equals the expected frequency caused by a combination of the thermodynamically limited gas generation rate and the intrinsic reaction rate for the consumption of water by radiolysis. By incorporating a simple gas-liquid diffusion model, the perturbation analysis indirectly confirms that the hydrogen gas pressure controls the rate of radiolytic gas generation in the water adsorbed on the plutonium oxide.

References

1. DOE Standard 3013: Stabilization, Packaging, and Storage of Plutonium-Bearing Materials, 2018.
2. J. M. Duffey and R. R. Livingston, "Gas Generation Testing of Plutonium Dioxide," Fifth Topical Meeting on Spent Nuclear Fuel and Fissile Materials Management, Charleston, South Carolina, September 17-20, 2002, WSRC-MS-2002-00705, September 17, 2002.
3. J. T. Pegg, A. E. Shields, M. T. Storr, D. O. Scanlon, and N. H. de Leeuw, "Interaction of Hydrogen with Actinide Dioxide Surfaces," J. Chem. Phys., 150, 134701, 2019.
4. H. E. Sims, K. J. Webb, J. Brown, D. Morris, and R. J. Taylor, "Hydrogen Yields from Water on the Surface of Plutonium Dioxide," J. Nucl. Mater., 437, 359-364, 2013.
5. D. K. Veirs, Los Alamos National Laboratory, private communication.
6. J. M. Duffey, Savannah River National Laboratory, private communication.
7. W. H. Press, S. A. Teukolsky, W. T. Vetterling, and B. P. Flannery, Numerical Recipes in FORTRAN: The Art of Scientific Computing, 2nd ed., Cambridge University Press, Cambridge, United Kingdom, 1992, pp. 569-577.
8. N. R. Lomb, "Least-Squares Frequency Analysis of Unequally Spaced Data," Astrophys. Space Sci., 39(2), 447-462, 1976.
9. CRC Handbook of Chemistry and Physics, 75th ed., CRC Press, Boca Raton, Florida (1994).
10. H. Christensen and S. Sunder, "Current State of Knowledge of Water Radiolysis Effects on Spent Nuclear Fuel Corrosion," Nucl. Technol., 131, 102-123, 2000.

608009 Photodecomposition of Organoiodides to Molecular Iodine As Pretreatment for Adsorption in Used Nuclear Fuel Recycling Operations

Abstract:

Capture and immobilization of volatile radionuclides such as ^{129}I, ^{85}Kr, ^{3}H, and ^{14}C resulting from used nuclear fuel (UNF) recycling operations is an essential component of the advanced fuel cycle. Particular attention is given to radioiodine due to its high mobility, half-life greater than 15 million years, and tendency to bio-accumulate. The dissolver off-gas (DOG) and the vessel off-gas (VOG) streams account for the majority of the radioiodine emissions with the VOG expected to contain iodine at parts per billion (ppb) levels. The chemical speciation of radioiodine significantly affects its capture by sorption, which is the preferred treatment technology. Organic iodide species are substantially more challenging to remove by adsorption compared to molecular iodine. We therefore demonstrate that by first decomposing organic iodide to molecular iodine and organics before sending to an adsorption column, the challenges can be significantly lessened. Specifically, we are investigating UV photolysis of methyl iodide in typical VOG conditions as a pretreatment step to adsorption.

Novel photochemical reactors for the UV photodecomposition of gaseous methyl iodide are presented in this work. The reactors are continuous and constructed of fluorinated ethylene propylene to allow transmission of UV light into the reactor while also providing resistance to the corrosive nature of methyl iodide. Integration of a titania nanotube photocatalyst increases reaction rate by two orders of magnitude compared to that of photolysis of methyl iodide with UV light alone. The UV light source is a low-pressure mercury vapor lamp and has a primary emission at 254 nm, a wavelength in the range of highest cross-sectional absorbance for methyl iodide. The titania photocatalyst is significantly active at low light intensity where without catalyst no detectable reaction occurs. For example, at an initial concentration of 400 ppb methyl iodide, a residence time of 9 s, and a light intensity of 0.5 mW/cm^2, conversions with and without titania photocatalyst are >99% and 0%, respectively.

Parametric analysis was carried out based on light intensity (UV lamp distance), gas composition (N_2 only vs air), and moisture. The titania photocatalyst was largely inactive without the presence of oxygen from air. Titania powder was shown to also significantly increase reaction rates compared to reaction without catalyst, however titania nanotubes produce higher reaction rates and offer easier methods for mechanically supporting the catalyst and integrating with the tubular reactors. The photolysis is affected by moisture, with reaction rates decreasing by a factor of six when the stream has 50% relative humidity compared to dry stream regardless of UV light intensity. The study performed demonstrates methyl iodide presence at ppb concentrations in VOG streams can effectively be completely and continuously decomposed with a common photocatalytic system and simple coil tube reactor. This supports the notion that VOG streams can be catalytically pretreated to facilitate the capture of radioiodine by downstream sorption systems.

600745 Thermodynamic Properties of (N,K)-a-S-H Gels for Radioactive Waste Solidification

Abstract:

Radioactive waste streams are often immobilised in Portland cement grouts to limit radionuclide leaching. Due to poor compatibility of certain wastes with Portland cement, alternatives such as alkali aluminosilicate 'geopolymers' are being investigated. Here, we describe the solubility and thermodynamic properties of synthetic sodium aluminosilicate hydrate gels (N-A-S-H) gels as the main reaction producing forming in metakaolin-based geopolymers. The solubility of solid N-A-S-H gels was determined by dissolution. The activities of the dissolved ionic species measured by ICP-OES were calculated using the Gibbs energy minimisation software GEMS v3.3 and the extended Debye-Hückel model. The aqueous and solid phases were characterised using X-ray diffraction, solid state ^{27}Al and ^{29}Si MAS and ^{1}H-^{29}Si CP MAS NMR and SEM-EDX. The (N,K)-A-S-H gels formed in the presence of trace amounts of Sr or Ca were a fully polymerised Al-rich (N,K)-A-S-H gels. Faujasite-Na and partially Sr-substituted zeolite Na-A form within the gels cured at 80 °C. Incorporation of Sr^{2+} or Ca^{2+} displaces some Na^{+} and K^{+} from the charge-balancing sites, with a slight decrease in the Si/Al ratio of the (N,K)-A-S-H gel. These results suggest that geopolymer gels are excellent candidates for immobilisation of radioactive waste containing alkaline earth radionuclides.

616084 Safety Document Development Process for DOE Nuclear Facilities

Abstract:

This presentation provides an overview of safety applications related to nuclear processes. Topics of interest that may be covered include a general overview of nuclear process safety, essential procedures and documents to perform a thorough safety analysis including criticality safety, and common process flammable gas safety evaluations.

616085 The Disruption of Vessel-Spanning Bubbles in Containers with Sloped Fins

Abstract:

Radioactive sludge was generated in the K-East Basin and K-West Basin fuel storage pools at the Hanford Site while irradiated uranium metal fuel elements from the N Reactor were being stored and packaged. The fuel has been removed from the K Basins. The sludge has been retrieved and loaded into sludge transport and storage containers (STSCs) and transported to T Plant for interim storage. The sludge is composed of a variety of particulate materials and water, including a fraction of reactive uranium metal particles that are a source of hydrogen gas. If a situation occurs where the reactive uranium metal particles settle out at the bottom of a container, previous studies have shown that a vessel-spanning gas layer above the uranium metal particles can develop and can push the overlying layer of sludge upward. The major concern, in addition to the general concern associated with the retention and release of a flammable gas such as hydrogen, is that if a vessel-spanning bubble (VSB) forms in an STSC, it may drive the overlying sludge material to the vents at the top of the container.

A previous study demonstrated that sloped walls on vessels provided an effective approach for disrupting a VSB by creating a release path for gas as a VSB began to rise. Based on the success of sloped-wall vessels, a similar concept is demonstrated here where a sloped fin is placed inside the vessel to create a release path for gas. A key potential advantage of using a sloped fin compared to a vessel with a sloped wall is that a small fin decreases the volume of a vessel available for sludge storage by a very small fraction compared to a cone-shaped vessel.

Experiments were conducted in 5, 10, and 23 in. diameter scaled vessels with a range of simulant shear strengths and fin shapes to determine what fin slope and width were sufficient to disrupt VSBs. Modeling of VSB stability and disruption were also used to support scaling of the results based on the gravity yield parameter ($Y_G = \tau_s/[\rho_s\ g\ D]$, where τ_s is the shear strength and ρ_s is the density of the simulant, D is the vessel diameter, and g is gravitational acceleration). Results for vessels of different size confirmed scaling with Y_G, though VSBs were shown to be more readily disrupted in progressively larger test vessels, which demonstrated that sloped fins would disrupt VSBs in the larger full-scale STSCs. A T-shaped fin with a 5-degree slope was selected for use in the full-scale STSCs and 20 STSCs have been loaded with radioactive waste and moved to a temporary storage location.

616086 Spray Releases from Pressurized Piping in Nuclear Facilities

Abstract:

An aerosol release from a leak or process piping failure is an important accident scenario to consider when performing nuclear facility safety analyses. Across the Department of Energy complex, DOE-HDBK-3010-94 has long been the source for acceptable methodologies to analyze various postulated events involving aerosol releases. It recommends bounding values of release fraction and airborne release fraction to be used to determine the source term from various phenomenologies. A recent independent technical review of DOE-HDBK-3010-94 identified some questions regarding the conservatism inherent in its recommended approaches. Concurrent with that review, Pacific Northwest National Laboratory (PNNL) embarked on a multi-year, multi-scale test program to measure aerosol releases from pressurized piping to support accident analyzes for the Waste Treatment and Immobilization Plant (WTP) on the Hanford site. The program examined the impact of an array of physical parameters that play a role in aerosol formation and transportation when a spray leak or release occurs. The test approach and key results from the PNNL test program will be discussed within the context of informing facility safety analyses.

614288 Fukushima-Daiichi: Challenges, Solutions, and Current Status

Abstract:

This keynote presentation discusses the Fukushima nuclear accident in 2011 that occurred due to damage from the tsunami following the Great East Japan Earthquake. The accident led to an air-borne release of nuclear material causing contamination of a large portion of the lands, and necessitating evacuation of the population. U.S. involvement (National Laboratories, EPA, etc.) to provide expertise in remediation technologies are discussed. The current status of the decontamination effort will be presented.

AUTHOR INDEX